Spring快速入门

崔彦威　卢欣欣　王倩　著

清华大学出版社
北京

内容简介

SSM目前是J2EE开发最常用、最流行的框架。本书将对Spring、Spring MVC、MyBatis、Spring Boot、Docker的使用进行介绍，每一章都有代码示例，便于理解每个章节的知识点，让读者掌握SSM框架，快速上手。

全书分为11章，内容包括Spring框架用到的注解、反射，Spring框架基础，核心容器，AOP，DAO，ORM与MyBatis，Spring MVC，Spring Boot配置，Spring Boot应用，Docker入门知识，最后实现一个用户权限管理的项目案例。

本书既适合Spring框架初学者学习（需要有Java编程基础），也适合高等院校和培训学校相关专业的师生参考。

图书在版编目（CIP）数据

Spring快速入门/崔彦威，卢欣欣，王倩著. —北京：清华大学出版社，2019

ISBN 978-7-302-53082-4

Ⅰ. ①S… Ⅱ. ①崔… ②卢… ③王… Ⅲ. ①JAVA语言—程序设计 Ⅳ. ①TP312.8

中国版本图书馆CIP数据核字（2019）第102171号

责任编辑：夏毓彦
封面设计：王　翔
责任校对：闫秀华
责任印制：沈　露

出版发行：清华大学出版社
网　　址：http://www.tup.com.cn，http://www.wqbook.com
地　　址：北京清华大学学研大厦A座　　**邮　　编：**100084
社 总 机：010-62770175　　**邮　　购：**010-62786544
投稿与读者服务：010-62776969，c-service@tup.tsinghua.edu.cn
质量反馈：010-62772015，zhiliang@tup.tsinghua.edu.cn
印 刷 者：北京鑫丰华彩印有限公司
装 订 者：三河市溧源装订厂
经　　销：全国新华书店
开　　本：190mm×260mm　　**印　　张：**18　　**字　　数：**461千字
版　　次：2019年7月第1版　　**印　　次：**2019年7月第1次印刷
定　　价：69.00元

产品编号：081773-01

前　言

平时我有写博客的习惯，喜欢将学到的知识点放在博客上：一是当作自己的学习笔记，将学习的内容整理之后再输出，也能够加深印象，忘记知识点时还可以快速复习；二是分享给有需要的朋友，希望各位在学习时能少走些弯路，少跳些坑。作为经常从网上索取免费资料的一员，我也要有回报的思想。

SSM 框架集目前是 J2EE 开发最常用、最流行的框架。Spring Boot 是由 Pivotal 团队提供的全新框架，设计目的是简化新 Spring 应用的初始搭建以及开发过程。Docker 容器技术在现在流行的 Devops 流水线上也扮演着重要的角色。

在本书中，我们将对 Spring、Spring MVC、MyBatis、Spring Boot、Docker 的使用进行介绍，而且每个章节基本都有代码示例，基本都是与技术相关、业务相关的，例子接近生活，便于读者对每个章节的知识点加深理解，快速上手。

本书读者对象

- 熟悉面向对象编程、经验丰富又打算学习 SSM、Spring Boot 的其他语言从业者。
- 有意提升网站和 Web 应用程序开发能力的 Web 开发人员。
- 希望在学习完 Java 编程想进一步提高开发技能的初学者。

阅读本书需要掌握 Java 面向对象编程知识，了解面向对象思想。

本书内容

本书共包括 11 章。第 1 章介绍 Java 基础，主要介绍 Spring 框架中常用的反射和注解技术，了解反射、注解相关概念。第 2 章先对 Spring 框架进行简单介绍，讲解 Spring 框架重要的 IOC、AOP 思想。第 3 章讲解 Spring 核心容器，介绍 Bean 的配置、注入方式、作用域和生命周期。第 4 章对 AOP 进行详细介绍，了解 AspectJ 的使用。第 5 章介绍 Spring 的 DAO 模块，同时了解 JDBC 的使用。第 6 章学习 MyBatis 的使用，主要包括 XML 的配置和映射，动态 SQL、逆向工程和 Pagehelper 的使用。第 7 章介绍 SSM 框架中的 SpringMVC，了解 Spring MVC 的处理流程、View 与 Controller 之间的数据传递。第 8、9 章主要介绍 Spring Boot 的相关知识以及 Spring Boot 配置，使用 Spring Boot 引入 Thymeleaf、JSP、MyBatis、Redis、Druid 等工具。第 10 章讲述 Docker 基础知识以及 Docker 的三大核心概念，并在 Docker 中使用 Tomcat 部署 war 包。第 11 章给出了一个项目实例，对前面章节介绍的知识点进行巩固。

本书导读

学习编程步骤可以用“学、练、悟、通”4个字概括。

（1）“学”指的是接收的过程，侧重理论。本书每个章节基本都是先介绍理论知识，让读者理解知识点为什么出现、要解决什么问题、有哪些优势。

（2）“练”指的是实践的过程。没有实践只有理论属于纸上谈兵，看的时候理解，操作时无从下手，动手能力差。本书每个章节都有实例，在学习理论的过程中可以参考实例操作一遍。

（3）“悟”指的是思考的过程。练更多的是模仿，照葫芦画瓢。在练的过程中也要多思考，多问几个为什么，多归纳总结，在做项目之前可以先把整个思路在脑子里过一遍。

（4）“通”指的是举一反三的过程。实现本书的例子不难，难的是将学到的知识举一反三，灵活地运用到其他项目中。可以找一些开源项目来研究，以达到融会贯通的境界。

示例源码下载

本书基本每个章节都有示例，完整源码可以扫描右边二维码，如果下载有问题，请联系技术支持邮箱 cuiyw525@163.com，邮件主题为“Spring 快速入门”。要运行本书中的示例，需要安装 Eclipse、Maven，并配置相关环境。

勘误与技术支持邮箱

作者已尽最大努力确保正文和代码没有问题。可是，金无足赤，疏漏在所难免。如果书中有错误，希望您能及时反馈给我们。我们将诚挚接受广大读者的批评指正，交流邮箱为 cuiyw525@163.com。勘误将发布在作者博客上：https://www.cnblogs.com/5ishare/。

致谢

出书、创业卖胡辣汤、做 IT 讲师是我大学时的梦想。这本书的出版也算是实现了我的一个梦想，当然实现梦想离不开大家的帮助。首先要感谢夏毓彦编辑，是他发现了我，给了我机会，并给予我自始至终的指导，还为我介绍同行朋友。感谢我的两个兄弟，陈岩亮、袁伟，大学时候的“铁三角”，我们一起度过了美好的大学生活。感谢卢欣欣、王倩、崔春英、陈立勇等教过我的大学老师，正是他们的教导让我对计算机产生了兴趣。感谢身边的同事和卢艳霞同学，他们给了我坚持写下去的动力。感谢生我养我的父母，长大后才慢慢理解了他们的不易，理解了生活的不易。最后感谢所有帮助过我的人。

本书封面照片由蜂鸟网的摄影家 ptwkzj 先生友情提供，在此表示衷心感谢。

崔彦威

2019 年 5 月于深圳

目　　录

第 1 章

Java基础

在学习 Spring 之前我们需要对 Java 基础语法有一定的了解，Java 中最重要的两个知识点是注解和反射。注解和反射在 Spring 框架中应用的最广泛。掌握注解和反射，有助于后面 Spring 的学习。

本章主要涉及的知识点：

- 注解基础：什么是注解?怎么理解注解?什么是元注解?
- 注解应用：自定义注解、注解的应用场景。
- 反射：反射的定义、反射的应用。

不管学习什么框架都需要先把 Java 基础夯实，基础打好之后才能厚积薄发。用到的时候不能只会用，不知道为什么这样用。学习编程还有最重要的一点就是需要勤动手，不能眼高手低，看着会做，真要动手时无从下手。

1.1 注解

本节首先介绍注解的基本概念，理解什么是注解、注解的作用是什么。在此基础上通过示例动手操作加深理解。

1.1.1 什么是注解

我们先看官方解释：它提供了一种安全的类似注释的机制，用来将任何的信息或元数据（metadata）与程序元素（类、方法、成员变量等）进行关联。为程序的元素（类、方法、成员变量）加上更直观、更明了的说明，这些说明信息与程序的业务逻辑无关，并且供指定的工具或框架使用。Annontation 像一种修饰符一样，应用于包、类型、构造方法、方法、成员变量、参数及本地变量的声明语句中。Java 注解是附加在代码中的一些元信息，便于一些工具在编译、运行时进行解析和使用，起到说明、配置的功能。注解不会也不能影响代码的实际逻辑，仅仅起到辅助性的作用，包含在 java.lang.annotation 包中。

看着上面的解释是不是还是一头雾水？其实我们可以更通俗地理解一下。最近几年出现一个词“斜杠青年”，还有黄某某拍摄的广告语：给人贴标签、下定义，总是很容易，而我却不会因为一件事被定性。这里的斜杠青年、贴标签都是把某些属性附加给对象，和注解功能差不多，它提供了一种安全的类似注释的机制，用来将任何信息或元数据（metadata）与程序元素（类、方法、成员变量等）进行关联。我们可以再来理解一下这句话，这里的程序元素可以理解为人，信息或元数据理解为标签，把标签属性（信息或元数据）赋给人（程序元素）。

上面两段基本把什么注解解释出来了，如果还是不知道注解是什么，那也没关系。其实我们在编程中已经用到或者看到过了，比如@Override、@Deprecated。是不是很熟悉？其实它们就是注解。

1.1.2 内置注解

上面的@Override、@Deprecated 都是 Java 中内置的注解，除了这两个还有其他的内置注解。这里列举了几个常用的内置注解以及它们的作用。

- @Deprecated：编译器在编译阶段遇到这个注解时会发出提醒警告，告诉开发者正在调用一个过时的元素，比如过时的方法、过时的类、过时的成员变量。
- @Override：提示子类要复写父类中被@Override 修饰的方法。
- @SuppressWarnings：阻止警告的意思。调用被@Deprecated 注解的方法后，编译器会警告提醒，而有时候开发者会忽略这种警告，他们可以在调用的地方通过@SuppressWarnings：达到目的。
- @SafeVarargs：参数安全类型注解。它的目的是提醒开发者不要用参数做一些不安全的操作，它的存在会阻止编译器产生 unchecked 这样的警告。它是在 Java 1.7 的版本中加入的。
- @FunctionalInterface：函数式接口注解，这个是 Java 1.8 版本引入的新特性。函数式编程很火，所以 Java 8 也及时添加了这个特性。函数式接口（Functional Interface）就是一个具有一个方法的普通接口。

1.1.3 元注解

通过前面的两小节，我们应该对注解有了一定的认识，下面进一步地了解一下注解。我们在自定义注解时会出现图 1-1 所示的一些选项。

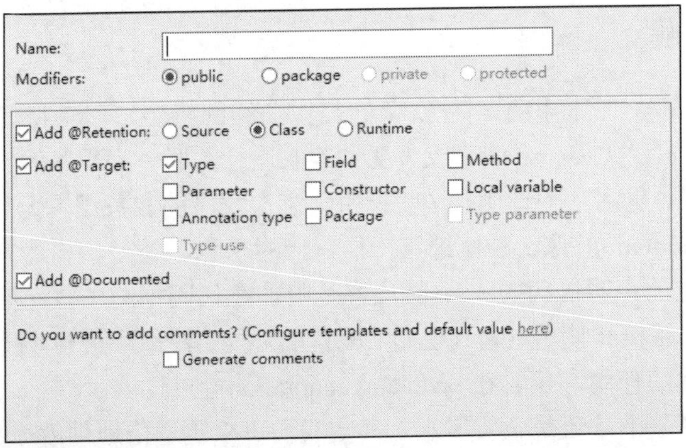

图 1-1

这些选项@Retention、@Target、@Documented 其实就是元注解。在创建时配置这些元注解，我们也可以推断出元注解的作用是什么。元注解负责注解自定义注解。java.lang.annotation 提供了 5 种元注解，专门注解其他的注解：

- @Retention：什么时候使用该注解。
- @Target：注解用于什么地方。
- @Documented：注解是否将包含在 JavaDoc 中。
- @Inherited：是否允许子类继承该注解。
- @Repeatable：指定注解可重复使用。

1. @Retention 定义注解的生命周期

- RetentionPolicy.SOURCE：在编译阶段丢弃。这些注解在编译结束之后不再有任何意义，所以它们不会写入字节码。@Override 和@SuppressWarnings 都属于这类注解。
- RetentionPolicy.CLASS：在类加载的时候丢弃。在字节码文件的处理中有用。注解默认使用这种方式。
- RetentionPolicy.RUNTIME：始终不会丢弃，运行期也保留该注解，因此可以使用反射机制读取该注解的信息。我们自定义的注解通常使用这种方式。

2. @Target 表示注解用于什么地方

默认值为任何元素，表示该注解用于什么地方。可用的 ElementType 参数包括：

- ElementType.CONSTRUCTOR：用于描述构造器。
- ElementType.FIELD：成员变量、对象、属性（包括 enum 实例）。
- ElementType.LOCAL_VARIABLE：用于描述局部变量。
- ElementType.METHOD：用于描述方法。
- ElementType.PACKAGE：用于描述包。
- ElementType.PARAMETER：用于描述参数。
- ElementType.TYPE：用于描述类、接口（包括注解类型）或 enum 声明。

3. @Documented 是一个简单的 Annotations 标记注解

表示是否将注解信息添加在 Java 文档中。

4. @Inherited 定义注解和子类的关系

@Inherited 元注解是一个标记注解，阐述了某个被标注的类型是被继承的。如果一个使用了@Inherited 修饰的 annotation 类型被用于一个 class，那么这个 annotation 将被用于该 class 的子类。

5. @Repeatable 指定注解可重复使用

使用@Repeatable 修饰表示该注解可以为重复使用。

1.1.4 自定义注解

元注解是负责注解自定义注解的。自定义注解时是有一些规则限制的，具体如下：

- Annotation 型定义为@interface，所有的 Annotation 会自动继承 java.lang.Annotation 这一接口，并且不能再去继承别的类或是接口。
- 参数成员只能用 public 或默认（default）这两个访问权修饰。
- 参数成员只能用基本类型 byte、short、char、int、long、float、double、boolean 八种基本数据类型和 String、Enum、Class、annotations 等数据类型，以及这一些类型的数组。
- 要获取类方法和字段的注解信息，必须通过 Java 的反射技术来获取 Annotation 对象，因为除此之外没有其他获取注解对象的方法。
- 注解也可以没有定义成员。

我们这里自定义一个注解来练习一下，主要用来演示自定义注解以及注解的继承。

1. 定义 CustomDescription 注解

CustomDescription 注解相当于标签。为了能多贴标签，又定义了注解容器 CustomDescriptions。其中，@Retention(RUNTIME)表示在运行时环境也可以获取注解，@Inherited 表示可继承，@Repeatable(CustomDescriptions.class)表示该注解可多次使用。

```
package CusAnnontation;

import static java.lang.annotation.ElementType.TYPE;
import static java.lang.annotation.RetentionPolicy.RUNTIME;
import java.lang.annotation.Documented;
import java.lang.annotation.Inherited;
import java.lang.annotation.Retention;
import java.lang.annotation.Target;
import java.lang.annotation.Repeatable;

@Documented
@Retention(RUNTIME)
@Target(TYPE)
@Inherited
@Repeatable(CustomDescriptions.class)
public @interface CustomDescription {
   String description() default "";
}
```

CustomDescriptions 容器：

```
package CusAnnontation;

import static java.lang.annotation.ElementType.TYPE;
import static java.lang.annotation.RetentionPolicy.RUNTIME;
import java.lang.annotation.Documented;
import java.lang.annotation.Inherited;
import java.lang.annotation.Retention;
import java.lang.annotation.Target;
```

```
@Documented
@Retention(RUNTIME)
@Target(TYPE)
@Inherited
public @interface CustomDescriptions {
   CustomDescription[] value();
}
```

2. 实现继承关系

这里为了演示，我们创建了两个类：一个基类 Person，一个子类 Student。在 Person 类加两个自定义注解，在 Student 中加一个自定义注解。

Person：

```
package CusAnnontation;

@CustomDescription(description="基类")
@CustomDescription(description="人")
public class Person {

   private String Name;

   public String getName() {
       return Name;
   }

   public void setName(String name) {
       Name = name;
   }
}
```

Student：

```
package CusAnnontation;

@CustomDescription(description="学生")
public class Student extends Person {
   private String StudentId;

   public String getStudentId() {
       return StudentId;
   }

   public void setStudentId(String studentId) {
       StudentId = studentId;
   }
}
```

3. 通过反射获取注解属性值

```
public static void main(String[] args) {
    // TODO Auto-generated method stub
    CustomDescriptions  customDescriptions =new Student().getClass().
getAnnotation(CustomDescriptions.class);
    for(CustomDescription h: customDescriptions.value()){
        System.out.println("description:" + h.description());
    }
}
```

这里我们想通过反射（可以先不要理解）获取 Student 类的注解值，那么问题来了，它是输出什么的呢？会输出“description:学生”吗？并不是，而是输出父类 Person 的注解。

输出：

```
description:基类
description:人
```

如果想输出子类 Student 的注解该怎么设置呢?很简单，只需在子类 Student 上覆盖父类的注解就好。

```
@CustomDescription(description="学生")
@CustomDescription(description="人")
public class Student extends Person
```

输出：

```
description:学生
description:人
```

此时输出的就是子类的注解值了。这里我们还可以验证@Retention 生命周期的作用，只需要把@Retention(RUNTIME)改成 CLASS，再运行就会报错，因为 main 方法中的 custormDescriptions 对象是一个 null 空值。不过自定义注解一般来说都是使用@Retention(RUNTIME)。

1.1.5 注解使用场景介绍

在上一小节通过实例学习了自定义注解的使用，之后就该解决怎么用的问题了。其实，注解应用的场景还是挺多的。

（1）使用注解做 bean 的属性值校验，例如在开发 Java 服务器端代码时，会要求对外部传来的参数合法性进行验证。hibernate-validator 提供了一些常用的参数校验注解。

（2）使用注解做权限控制。例如，shiro 框架中有 5 个权限注解，我们也可以自定义注解进行权限控制。

（3）代替配置文件功能，像 Spring 基于注解的配置，减少了 xml 的配置。

（4）可以生成文档，像 Java 代码注释中的@see、@param 等。

这里只是列举了几个使用场景，其实还有很多地方可以使用注解。

1.2 反射

本节首先介绍反射的基本概念，理解什么是反射，以及 Class 类和反射常用 API，通过实例操作来学习反射的使用。

1.2.1 反射机制

在上面自定义注解时我们也有提到反射，要获取类方法和字段的注解信息，必须通过 Java 的反射技术来获取 Annotation 对象。那么什么是反射呢？在运行状态中，对于任意一个类，都能够知道这个类的所有属性和方法；对于任意一个对象，都能够调用它的任意一个方法和属性，这种动态获取的信息以及动态调用对象的方法的功能称为 Java 语言的反射机制。它有点类似照妖镜的作用，不管是什么妖魔鬼怪（类或对象）都能看到它的真面目（获取类的属性方法、调用对象的属性方法）。

1.2.2 理解 Class 类

反射机制可以动态获取类信息以及调用对象方法，那它是通过什么实现的呢？这就要介绍一下 Class 类了。首先明确 Class 也是一个类，只是它是一个描述类的类，也可以生成对象。对于每个类而言，在 JRE 中有且仅有一个不变的 Class 类型的对象，而这个 Class 类型的对象只能由系统建立，封装了当前对象所对应的类的信息，有哪些属性、方法、构造器以及实现了哪些接口等。每个类的实例都会记得自己是由哪个 Class 实例所生成的。

要获取类信息或调用对象方法，肯定首先要获取到该类或对象对应的 Class 类的实例。一般获取 Class 对象有 3 种方式。

- 通过类名获取，类名.class。
- 通过对象获取，对象.getClass()。
- 通过全类名获取，Class.forName（全类名）。

这里我们可以使用字符串来做验证。

```
public static void main(String[] args) throws ClassNotFoundException {
    //字符串的例子
    Class clazz = null;
    //类名.class
    clazz = String.class;
    System.out.println(clazz);
    //对象.getClass()
    clazz = "ReflectionTest".getClass();
    System.out.println(clazz);
    //Class.forName(全类名)
```

```
    clazz = Class.forName("java.lang.String");
    System.out.println(clazz);

}
```

输出结果：

```
class java.lang.String
class java.lang.String
class java.lang.String
```

通过 3 种方式获取到 Class 实例后，再了解一下 Class 类常用的方法（见表 1-1）。

表 1-1　Class 类常用的方法

方法名	功能说明
forName(String name)	返回指定类名 name 的 Class 对象
newInstance()	调用默认构造函数，返回该 Class 对象的一个实例
newInstance(Object []args)	调用当前格式构造函数，返回该 Class 对象的一个实例
getName()	返回此 Class 对象所表示的实体（类、接口、数组类、基本类型或 void）名称
getSuperClass()	返回当前 Class 对象的父类的 Class 对象
getInterfaces()	获取当前 Class 对象的接口
getClassLoader()	返回该类的类加载器
getSuperclass()	返回表示此 Class 所表示的实体的超类的 Class
getFields()	获取类中 public 类型的属性
getField(String name)	获取类特定的方法，name 参数指定了属性的名称
getDeclaredFields()	获取类中所有的属性（public、protected、default、private），但不包括继承的属性
getDeclaredField(String name)	获取类特定的方法，name 参数指定了属性的名称
getConstructors()	获取类中的公共方法
getConstructor(Class[] params)	获取类的特定构造方法，params 参数指定构造方法的参数类型
getDeclaredConstructors()	获取类中所有的构造方法（public、protected、default、private）
getDeclaredConstructor(Class[] params)	获取类的特定构造方法，params 参数指定构造方法的参数类型
getMethods()	获得类的 public 类型的方法
getMethod(String name, Class[] params)	获得类的特定方法，name 参数指定方法的名字，params 参数指定方法的参数类型
getDeclaredMethods()	获取类中所有的方法（public、protected、default、private）
getDeclaredMethod(String name, Class[] params)	获得类的特定方法，name 参数指定方法的名字，params 参数指定方法的参数类型

1.2.3 反射的使用

这里要着重介绍一下上面 API 的使用，因为在后面要学习的 Spring 中 IOC 的原理就是反射加工厂模式。学好反射 API 有助于理解 Spring 框架内部实现。为了演示 Class 方法的使用，在注解 demo 的基础上对 Person、Student 类进行了修改。

Person 类：

```
package Reflection;

@CustomDescription(description="基类")
@CustomDescription(description="人")
public class Person {

    private String Name;

    public String getName() {
        return Name;
    }

    public void setName(String name) {
        Name = name;
    }

    public String PersonPublicMethod(String str)
    {
        return str;
    }

    public Person(String name) {
        Name = name;
    }

    public String PersonPrivateMethod(String str)
    {
        return str;
    }

    public Person() {
        super();
    }
}
```

Student 类：

```
package Reflection;
```

```
@CustomDescription(description="学生")
@CustomDescription(description="人")
public class Student extends Person {
    public String StudentId;

    public String getStudentId() {
        return StudentId;
    }

    public void setStudentId(String studentId) {
        StudentId = studentId;
    }

    public String StudentPublicMethod(String str)
    {
        return str;
    }

    private String StudentPrivateMethod(String str)
    {
        return str;
    }

    public Student(String name, String studentId) {
        super(name);
        StudentId = studentId;
    }

    private Student(String name) {
        super(name);
        StudentId="123456";
    }

    public Student() {

    }
}
```

1. 描述方法 Method

描述方法主要是 4 个获取方法（getMethods、getMethod、getDeclaredMethods、getDeclaredMethod）和 1 个调用方法（invoke）。

- getMethods：获取 clazz 对应类中的所有方法，不能获取 private 方法，且获取从父类继承来的所有方法，包括私有父类的私有方法。
- getMethod：获取 clazz 对应类中指定方法名和参数类型的方法，不能获取 private 方法，且获取从父类继承来的所有方法，包括私有父类的私有方法。因为存在同方法名不同参数这种情况，所以只有同时指定方法名和参数类型才能唯一确定一个方法。

- getDeclaredMethods：获取所有方法，包括私有方法，所有声明的方法，都可以获取到，且只获取当前类的方法。
- getDeclaredMethod：获取 clazz 对应类中指定方法名和参数类型的方法，包括私有方法，所有声明的方法，都可以获取到，且只获取当前类的方法。
- invoke：执行方法，第一个参数表示执行哪个对象的方法，剩下的参数是执行方法时需要传入的参数，私有方法的执行必须在调用 invoke 之前加上一句“method.setAccessible(true);”。

```
Class clazz = Class.forName("Reflection.Student");
Method method=null;
Method[] methods=null;

methods = clazz.getMethods();
for(Method mth:methods){
    System.out.print(mth.getName()+" ");
}
System.out.println();

method = clazz.getMethod("StudentPublicMethod",String.class);
System.out.print(method.getName()+" ");
System.out.println();

methods = clazz.getDeclaredMethods();
for(Method mth:methods){
    System.out.print(mth.getName()+" ");
}
System.out.println();

method = clazz.getDeclaredMethod("StudentPrivateMethod", String.class);
System.out.print(method.getName()+" ");
System.out.println();

Object obje = clazz.newInstance();
method.setAccessible(true);
String result=(String) method.invoke(obje,"inputParams");
System.out.println(result);
```

输出结果：

```
<terminated> ReflectionTest [Java Application] C:\Program Files\Java\jre1.8.0_162\bin\javaw.exe (2018年8月10日 下午9:35:47
StudentPublicMethod setStudentId getStudentId getName setName PersonPrivateMethod PersonPublicMethod wa
StudentPublicMethod
StudentPrivateMethod StudentPublicMethod setStudentId getStudentId
StudentPrivateMethod
inputParams
```

上面我们基本可以实现通过类名创建对象、通过方法名执行方法。类名和方法名都是字符串，我们可以把它们放到一个配置文件中，根据配置文件来执行方法，这样就有点类似基于 XML 的 Spring 了。

2. 描述字段 Field

描述字段 Field 方法的使用和描述方法 Method 中方法的使用有点类似，也是 4 个获取字段的方法（getFields、getField、getDeclaredFields、getDeclaredField）。

- getFields：获得某个类的所有公共（public）字段，包括父类中的字段。
- getField：获取某个类 public 成员变量中指定变量名的字段，包括基类。
- getDeclaredFields：获得某个类所有声明的字段，包括 public、private 和 protected，但是不包括父类的声明字段。
- getDeclaredField：获取某个类的所有成员变量指定变量名的字段，不包括基类。

```
Class clazz = Class.forName("Reflection.Student");
System.out.println("---------getDeclaredFields---------");
Field[] fields = clazz.getDeclaredFields();
for(Field field: fields){
    System.out.print(field.getName()+" ");
}
System.out.println();
System.out.println("---------getFields---------");
fields = clazz.getFields();
for(Field field: fields){
     System.out.print(field.getName()+" ");
}
System.out.println();

System.out.println("---------getDeclaredField---------");
Field field = clazz.getDeclaredField("StudentId");
field.setAccessible(true);
System.out.println(field.getName());

System.out.println("---------getField--------");

field = clazz.getField("StudentId");
System.out.println(field.getName());
```

输出结果：

```
---------getDeclaredFields---------
StudentId
---------getFields---------
StudentId
---------getDeclaredField---------
StudentId
---------getField--------
StudentId
```

上面通过反射获取字段，得到字段之后就是获取或设置字段的值了。如果字段是私有的，那么不管是读值还是写值，都必须先调用 setAccessible(true)方法，比如在 Person 类中，字段 name 字段是私有的。

```
Class clazz = Class.forName("Reflection.Person");
Person person = new Person("CYW");
//获取私有字段的值
Field field = clazz.getDeclaredField("Name");
//由于是私有字段，因此需要使用 setAccessible(true)
field.setAccessible(true);
Object val = field.get(person);
System.out.println(val);
//改变私有字段的值
field.set(person, "ivan");
System.out.println(person.getName());
```

输出结果：

```
CYW
ivan
```

3. 描述构造器 Constructor

先介绍一下描述构造函数 Constructor 用到的方法，主要还是 4 个：getConstructors、getDeclaredConstructors、getConstructor、getDeclaredConstructor。和前面 Method、Field 用的方法进行比较，举一反三，我们也能大概了解这几个方法的使用。其实，在编程中有好多体现哲学思想的地方，有正有反，有阴有阳，学会思考，这样可以以点带面、触类旁通。

- getConstructors：获取对应类中 public 类型的构造函数，且只获取当前类的构造函数。
- getConstructor：获取对应类中 public 指定参数类型的构造函数，且只获取当前类的构造函数。
- getDeclaredConstructors：获取对应类中所有构造函数，包括私有构造函数，且只获取当前类的构造函数。
- getDeclaredConstructor：获取对应类中指定参数类型的方法，包括私有构造函数，且只获取当前类的方法。

```
String className = "Reflection.Student";
Class<Student> clazz = (Class<Student>) Class.forName(className);

//指定成父类之后实际还是获取子类的构造函数
Constructor<Person> [] constructors =
        (Constructor<Person>[]) Class.forName(className).
getConstructors();

for(Constructor<Person> constructor: constructors){
    System.out.println("getConstructors:"+constructor);
}
```

```
            Constructor<Student> []  constructorsa =
                    (Constructor<Student>[]) Class.forName(className).
getDeclaredConstructors();

            for(Constructor<Student> constructor: constructorsa){
                System.out.println("getDeclaredConstructors:"+constructor);
            }


        //通过 getConstructor 获取公有构造函数
            Constructor<Student> constructor =
clazz.getConstructor(String.class, String.class);
            System.out.println("getConstructor:"+constructor);
            Student obj = constructor.newInstance("cyw", "123456");
            System.out.println(obj.getName());

            //通过 getDeclaredConstructor 获取私有构造函数
            constructor = clazz.getDeclaredConstructor(String.class);
            System.out.println("getDeclaredConstructor:"+constructor);
            //对于私有构造函数在初始化之前要设置 setAccessible(true)
            constructor.setAccessible(true);
            obj = constructor.newInstance("cyw");
            System.out.println(obj.getName());
```

输出结果：

```
getConstructors:public Reflection.Student(java.lang.String,java.lang.String)
getConstructors:public Reflection.Student()
getDeclaredConstructors:public Reflection.Student(java.lang.String,java.lang.String)
getDeclaredConstructors:private Reflection.Student(java.lang.String)
getDeclaredConstructors:public Reflection.Student()
getConstructor:public Reflection.Student(java.lang.String,java.lang.String)
cyw
getDeclaredConstructor:private Reflection.Student(java.lang.String)
cyw
```

4. 描述注解 Annotation

描述注解主要用到 getAnnotation(Class<A> annotationClass)方法，返回该元素指定类型的注解，否则返回 null。

```
            String className = "Reflection.Student";
            Class<Student> clazz = (Class<Student>) Class.forName(className);
            CustomDescriptions  customDescriptions
 =clazz.getAnnotation(CustomDescriptions.class);
            for(CustomDescription h: customDescriptions.value()){
                System.out.println("description:" + h.description());
            }
```

输出结果：

```
description:学生
description:人
```

1.3 小结

我们回顾一下这一章的主要内容：在注解部分，学习了注解的定义、注解的理解、内置注解、元注解，通过自定义注解理解注解继承、多注解及注解生命周期；在反射部分，了解反射机制、Class 类的理解，通过实例掌握反射中 Method、Field、Constructor、Annotation 相关 API 的使用，为后续学习 Spring 框架打下基础。

第 2 章

Spring基础

在上一章节中，我们学习了 Java 的注解与反射，在这一章节我们将了解一下 Spring 框架，并学习 Spring 框架中重要的编程思想控制反转（IOC）、面向切面编程（AOP）。语言只是工具，最重要的是编程思想。掌握了编程思想，不仅是 Java，其他编程语言也就容易学习了。

本章主要涉及的知识点：

- Spring 概述：了解 Spring 框架的起源、简介、结构。
- 依赖注入：什么是依赖注入?依赖注入的好处是什么?
- 控制反转：什么是控制反转?依赖注入与控制反转的联系是什么?
- 面向切面编程：什么是面向切面编程?面向切面编程的原理。
- 实例应用：通过本章 IOC、AOP 示例，演示 Spring 中 IOC、AOP 的简单应用，通过动手实践加深对 IOC、AOP 的理解。

2.1 Spring 框架介绍

本节主要对 Spring 框架进行简单介绍，了解框架起源、框架组成结构，对 Spring 框架有一个大概的认识。

2.1.1 起源

首先，追根溯源，了解它是怎么来的。在 Spring 框架出现之前，使用 EJB 开发 J2EE 应用可没那么容易。EJB 要严格地实现各种不同类型的接口，代码复用性低，配置也比较复杂和单调，同样使用 JNDI 进行对象查找的代码也是单调而枯燥，而且 EJB 不容易学，开发效率低。Spring 出现的初衷就是为了解决类似的这些问题。

Spring 最大的目的之一就是使 J2EE 开发更加容易。同时，Spring 不仅仅是一个单层的框架，而是类似一个平台或者生态体系。在这个平台或者生态体系中，可以将 Struts、Hibernate 等单层框架最佳的方式融合在一起，为企业级应用提供完美的解决方案。Spring 的形成，最初来自 Rod Jahnson 所著的一本很有影响力的图书《Expert One-on-One J2EE Design and Development》（出版于 2002 年），就是在这本书中第一次出现了 Spring 的一些核心思想。另外，《Expert One-on-One J2EE Development without EJB》更进一步地阐述了在不使用 EJB 开发 J2EE 企业级应用的一些设计思想和具体的做法。

2.1.2　简介

了解了 Spring 框架的起源之后，下面来了解一下 Spring 到底是什么。用一句话概括，Spring 就是一个开源的轻量级的控制反转（IoC）和面向切面（AOP）的容器框架。下面来分析一下这句话。

（1）开源：因为开源、免费，用户无须经过任何人同意即可修改代码，可控制性强，不受他人限制。

（2）轻量级：从大小与开销两方面而言，Spring 都是轻量的。完整的 Spring 框架可以在一个大小只有 1MB 多的 JAR 文件里发布，并且 Spring 所需的处理开销也是微不足道的。此外，Spring 是非侵入式的，Spring 应用中的对象不依赖于 Spring 的特定类。

（3）控制反转：软件设计中通常用耦合度和内聚度作为衡量模块独立程度的标准，划分模块的一个准则就是高内聚低耦合，Spring 通过控制反转技术降低了耦合度。

（4）面向切片：Spring 支持面向切片的编程，允许通过分离应用的业务逻辑与系统级服务进行内聚性的开发，应用对象只需实现业务逻辑，它们并不负责（甚至是意识）其他系统级关注点，例如日志或事务支持。

（5）容器：容器就是用来装东西的。Spring 容器包含并管理应用对象的配置和生命周期。

（6）框架：Spring 可以将简单的组件配置、组合成为复杂的应用，相当于是一个脚手架，开发者要做的就是把组件放进去，实现业务逻辑。

2.1.3　框架结构

Spring 框架结构如图 2-1 所示。

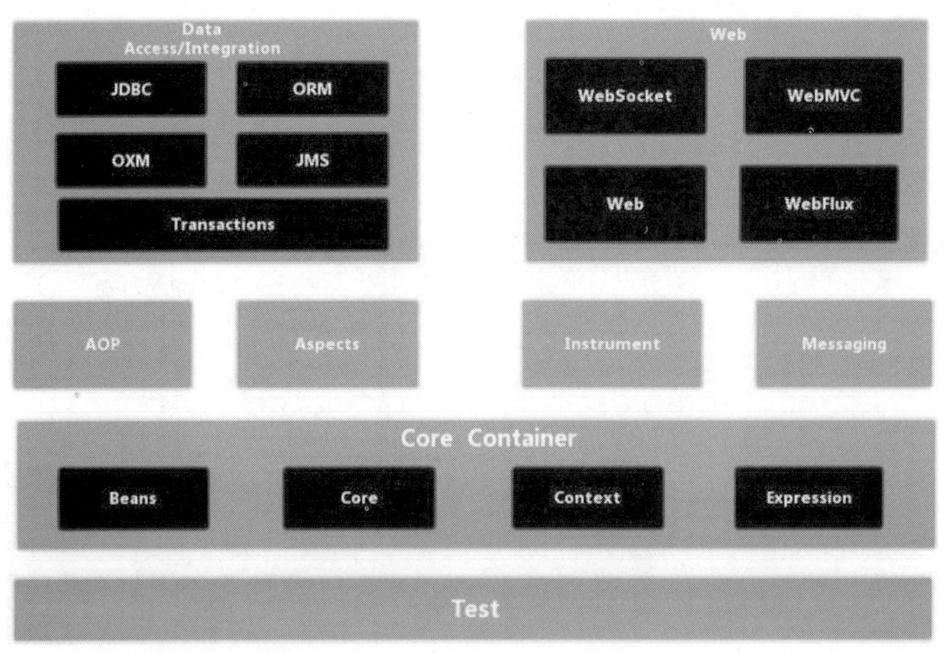

图 2-1

Spring 由 20 多个模块组成，可以分为核心容器（Core Container）、数据访问/集成（Data Access/Integration）、Web、面向切面编程（AOP，Aspect Oriented Programming）、设备（Instrumentation）、消息发送（Messaging）和测试（Test）。

1. 核心容器

核心容器包含 spring-core、spring-beans、spring-context、spring-context-support 和 spring-expression（Spring Expression Language）这些模块。

spring-core 和 spring-beans 构成了框架最基础的部分，包括控制反转和依赖注入功能。

spring-context 是基于 spring-core 和 spring-beans 构建的，提供了一种以框架风格来访问对象的方式，类似于 JNDI 注册。ApplicationContext 接口是 spring-context 的焦点。

spring-context-support 为集成第三方库（如定时器 Quartz）提供支持。

spring-expression 提供了一种强大的表达式语言，可以在运行时查询和操作对象。

2. AOP

spring-aop 模块提供了一个 AOP 面向切面编程的实现。

spring-aspects 模块提供与 AspectJ 的集成。

spring-instrument 模块提供一些类级的工具支持和 ClassLoader 级的实现，用于服务器。spring-instrument-tomcat 模块针对 tomcat 的 instrument 实现。

3. 消息发送

从 Spring 4 开始包含了一个 spring-messaging 模块，对 Spring 集成项目 Message、MessageChannel 和 MessageHandler 进行了重要的抽象，是基于消息发送应用的基础。

4. 数据访问/集成

数据访问/集成层包含 JDBC（spring-jdbc）、ORM（spring-orm）、OXM（spring-oxm）、JMS（spring-jms）和事务（spring-tx）模块。

5. Web

Web 层包含 spring-web、spring-webmvc、spring-websocket 和 spring-webflux 模块。其中，spring-web 提供了面向 Web 集成的基本特性，比如文件上传功能。Spring-webmvc 模块包含了 Spring 的 MVC 和 REST Web Service 实现。spring-webflux 是一个新的非堵塞函数式 Reactive Web 框架，可以用来建立异步的、非阻塞、事件驱动的服务，并且扩展性非常好。

6. 测试

Spring-test 模块支持 Spring 组建 JUnit 和 TestNG 的单元测试和集成测试。

2.2 依赖注入 DI 与控制反转 IOC

本节主要介绍依赖注入与控制反转的基本概念，理解什么是依赖注入、什么是控制反转。在此基础上通过示例动手操作加深理解。

2.2.1 什么是依赖注入

从汉字的字面意思理解，可以把“依赖注入”分为两个词，一个是“依赖”，一个是“注入”。那什么是依赖、什么是注入呢?依赖是依靠别人或事物而不能自立或自给，通俗的理解就是，不是自身的，但没有就活不下去，比如人没有了空气、水、阳光，那就活不下去，所以人依赖空气、水、阳光。注入是之前内部没有通过外部灌入的。

上面是从字面意思理解了一下依赖注入，接着从编程的角度分析一下。依赖注入是组件之间依赖关系由容器在运行期决定的，即由容器动态地将某个依赖关系注入到组件之中。依赖注入的目的并非为软件系统带来更多功能，而是为了解耦，提升组件重用的频率，并为系统搭建一个灵活、可扩展的平台。通过依赖注入机制，我们只需要通过简单的配置，而无须任何代码就可以指定目标需要的资源，完成自身的业务逻辑，不需要关心具体的资源来自何处、由谁实现。

- **依赖注入是组件之间依赖关系由容器在运行期决定的，即由容器动态地将某个依赖关系注入到组件之中。**这句话总结依赖注入的核心原理，比如分别把人（Person）和空气（Air）都当作一个组件（类），人依赖空气，那么这个依赖关系不是人和空气两个组件关联的，需要一个容器决定，而且不是一开始就决定的，是运行期决定的。
- **依赖注入的目的并非为软件系统带来更多功能，而是为了解耦，提升组件重用的频率，并为系统搭建一个灵活、可扩展的平台。**这句话是使用依赖注入的目的。使用依赖注入可以进一步解耦，也是软件设计中高内聚低耦合的体现。
- **通过依赖注入机制，我们只需要通过简单的配置，而无须任何代码就可以指定目标需要的资源，完成自身的业务逻辑，不需要关心具体的资源来自何处、由谁实现。**这句话是实现依赖注入的方法，通过简单配置就能实现依赖注入。

理解 DI 的关键是：“谁依赖谁，为什么需要依赖，谁注入谁，注入了什么”，接着使用人和空气来深入分析一下：

- 谁依赖谁：人依赖空气。
- 为什么需要依赖：人需要空气，没有空气，人就不存在，就是一个空对象。
- 谁注入谁：人通过容器注入空气。
- 注入了什么：注入了人所需要的空气。

2.2.2 什么是控制反转

与依赖注入一样，我们还是先从字面意思理解一下控制反转。它也可以分为两个词，一个是“控制”，一个是“反转”。那什么是控制，什么又是反转呢？

- 控制：为掌握住对象不使任意活动或超出范围；或使其按控制者的意愿活动。
- 反转：向相反的方向转动。

上面是从中文字面对控制反转的理解，下面从编程的角度分析一下。IOC 意味着将你设计好的对象交给容器控制，而不是传统的在你的对象内部直接控制。理解 IOC 的关键是："谁控制谁，控制什么，为何是反转（有反转就应该有正转了），哪些方面反转了"。我们接着使用人和空气来深入分析一下：

- 谁控制谁，控制什么：以往我们是通过 new 关键字来创建对象的，比如人（Person）依赖空气（Air），在 Person 中如果要使用 Air 对象，就需要通过 New 关键字来主动创建，在 IOC 中有一个专门的容器来创建这些对象，即由 IOC 容器来控制对象的创建。
  - 谁控制谁？当然是 IOC 容器控制对象了。
  - 控制什么？主要控制了外部资源获取（不只是对象，还包括文件等）。
- 为何是反转，哪些方面反转了：有反转就有正转，正转就是我们通过 New 主动获取依赖对象；反转则是由容器来帮忙创建及注入依赖对象。
  - 为何是反转？因为由容器帮我们查找及注入依赖对象，对象只是被动地接受依赖对象，所以是反转。
  - 哪些方面反转了？依赖对象的获取被反转了。Air 对象通过容器注入给 Person 对象。

2.2.3 依赖注入的优缺点

可能有人会纳闷为什么只介绍使用依赖注入的优缺点，那控制反转就没有优缺点吗？其实它们是同一个概念的不同角度描述，由于控制反转概念比较含糊（可能只是理解为容器控制对象这一个层面，很难让人想到谁来维护对象关系），所以 2004 年大师级人物 Martin Fowler 又给出了一个新的名字："依赖注入"。相对 IOC 而言，"依赖注入"明确描述了"被注入对象依赖 IOC 容器配置依赖对象"。

俗话说，每个硬币都有两面。同样，IOC 也是有优点和缺点的。

优点，也是使用依赖注入的目的：实现组件之间的解耦，提高程序的灵活性和可维护性，提升组件重用的频率，并为系统搭建一个灵活、可扩展的平台。

缺点也是存在的。一是它创建对象的方式变复杂了。二是因为使用反射来创建对象，所以在效率上会有些损耗。但相对于程序的灵活性和可维护性来说，这点损耗是微不足道的。三是使用 XML 进行配置时太复杂，一旦类有改变，XML 就需要改变。有了注解之后，与使用 XML 进行配置相比简单很多。

2.2.4 IOC 实例

光说不练假把式，特别是 IT 技术，经常会出现看能看懂但写不出来的尴尬局面。上面分析了一下依赖注入与控制反转，本节将通过示例来加深理解。这里还是使用人与空气的例子。

人依赖空气，在传统的方式创建两个类：一个是 Person 类，一个是 CleanAir 类。

```
public class CleanAir {
    @Override
```

```
    public String toString() {

        return  "CleanAir";
    }
}
public class Person {

    public Person(CleanAir air) {

        this.air = air;
    }
    CleanAir air;

    public void Breath()
    {
            System.out.print(this.air.toString());
    }
}
```

上面两个类实现了依赖的关系，还有就是注入。在了解注入之前，我们还有一个问题要思考。有这样一句话：世界上唯一不变的就是变化。之前干净的空气不复存在，而 Person 依赖的不再是 CleanAir，而是比 CleanAir 更有内涵的 DirtyAir。如果还是按照上面的方式来，那就需要在增加一个 DirtyAir 类的同时修改 Person。这种强依赖有很大的弊端，一个地方变化引起其他地方变化，而且改变的只是 Air，但 Person 也要改变。怎么样才能尽量减少修改的地方呢？于是面向接口的编程出现了。下面先定义一个接口 IAir，类 CleanAir 实现接口 IAir，在 Person 中不再直接依赖 CleanAir，而是依赖接口 IAir，这样即使是 DirtyAir 也只需要给 Person 修改不同的 Air 就行了。这个注入的过程，利用 Spring 框架只需要改一下配置即可实现。

```
public interface IAir {}

public class CleanAir implements IAir {
    @Override
    public String toString() {

        return  "CleanAir";
    }
}

public class Person {

    public Person(IAir air) {

        this.air = air;
    }
    IAir air;
```

```
    public void Breath()
    {
        System.out.print(this.air.toString());
    }
}

public class DirtyAir implements IAir {
    @Override
    public String toString() {

        return  "DirtyAir";
    }
}
```

上面定义 IAir 接口算是对依赖关系的优化，降低了人与空气的耦合度，但是并没有使用 New 关键字创建对象，只是定义了依赖关系。下面用 Spring 实现注入。

（1）创建一个 Maven Project，archetype 选择 quickstart，如图 2-2 所示。

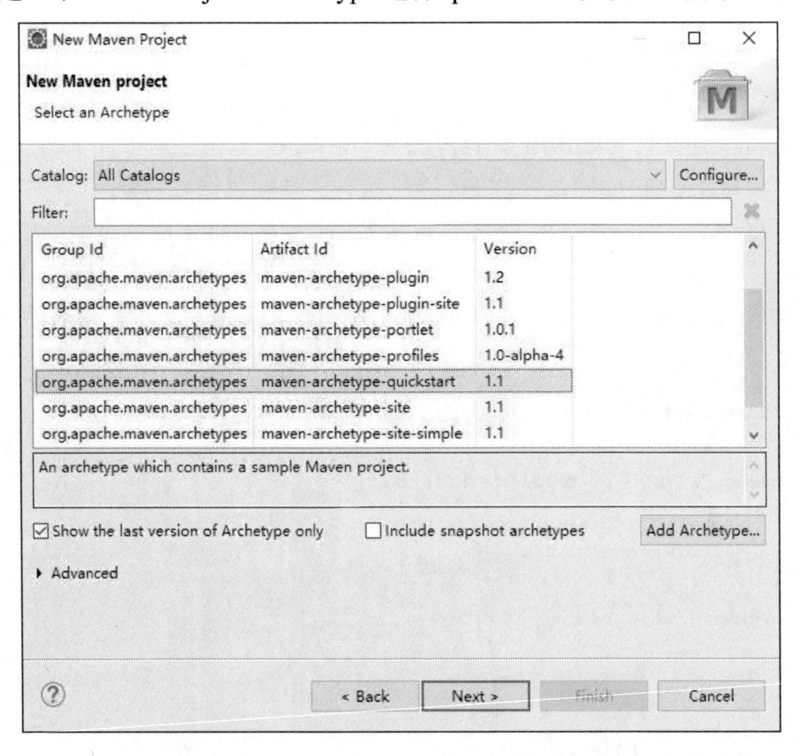

图 2-2

（2）创建之后，既然要使用 Spring 框架来实现注入，那肯定要在项目中引入 Spring 框架，配置 pom.xml，添加依赖。

① 在 properties 节点配置要引入 Spring 的版本号，这里用的是 5.0.0.RELEASE。

```
<spring.version>5.0.0.RELEASE</spring.version>
```

② 引入。

```
<dependency>
    <groupId>org.springframework</groupId>
    <artifactId>spring-core</artifactId>
    <version>${spring.version}</version>
</dependency>
<dependency>
    <groupId>org.springframework</groupId>
    <artifactId>spring-context</artifactId>
    <version>${spring.version}</version>
    </dependency>
```

（3）实现依赖注入。

前面创建的 IAir 接口、Person 类、CleanAir 类、DirtyAir 类实现了依赖关系，但是怎么让 Spring 框架识别到呢?不可能把各个类创建好就结束了，还需要进行配置才能让 Spring 知道哪些是组件。这里对 Person 类、CleanAir 类、DirtyAir 类在上面定义的基础上进行修改。

```
@Component
public class CleanAir implements IAir {
    @Override
    public String toString() {

        return  "CleanAir";
    }
}

@Component
public class DirtyAir implements IAir {
    @Override
    public String toString() {

        return  "DirtyAir";
    }
}

@Component
public class Person {

@Autowired //对构造函数进行标注

    public Person(@Qualifier("dirtyair")IAir air) {

        this.air = air;
    }

    IAir air;
```

```
    public void Breath()
    {
        System.out.print(this.air.toString());
    }
}
```

在上面的代码中使用了@Component将类注解成组件，使用@Autowired将IAir类型对象注入Person中。CleanAir类、DirtyAir类都实现了IAir接口，怎么让Person具体注入哪个对象呢？使用@Qualifier关键字来进行区分，这里使用的是qualifier=dirtyair。同时这些组件定义之后还要告诉Spring框架组件位置在哪，所以在scr/main/resources下新建了ApplicationContext.xml进行配置。

```
<?xml version="1.0" encoding="UTF-8"?>
<beans xmlns="http://www.springframework.org/schema/beans"
    xmlns:xsi="http://www.w3.org/2001/XMLSchema-instance"
    xmlns:context="http://www.springframework.org/schema/context"
    xsi:schemaLocation="http://www.springframework.org/schema/beans
        http://www.springframework.org/schema/beans/spring-beans.xsd
        http://www.springframework.org/schema/context
        http://www.springframework.org/schema/context/spring-context.xsd"
>

<context:component-scan base-package="com.demo.model"/>
    <bean id="CleanAir" class="com.demo.model.CleanAir">
        <qualifier value="cleanair"/>
    </bean>
    <bean id="DirtyAir" class="com.demo.model.DirtyAir">
        <qualifier value="dirtyair"/>
    </bean>
    <bean id="person" class="com.demo.model.Person"/>
</beans>
```

（4）测试。

在main中获取到应用上下文，通过getBean方法获取Person对象，然后调用Breath()方法。不了解ClassPathXmlApplicationContext、getBean这些方法也没关系，在后面的章节会有详细介绍。

```
    public static void main( String[] args )
    {
        ApplicationContext context=new ClassPathXmlApplicationContext(new
String[]{"ApplicationContext.xml"});
        BeanFactory factory=context;
        Person person=(Person)factory.getBean("person");
        person.Breath();

    }
```

输出结果：

```
DirtyAir
```

如果想使用 CleanAir 对象，只需要把 Person 类中@Qualifier 注解 value 的值改为 CleanAir 对应的 beanId：cleanair。

（5）小结

参考上面的示例思考 2.2.1 和 2.2.2 节中的概念，就会发现依赖注入、控制反转其实也不难。理解了依赖注入、控制反转对后面 Spring 框架的学习会有更大的帮助。

2.3 面向切面编程

本节主要认识横切、纵切，理解什么是 AOP（Aspect-Oriented Programming，面向切面编程）以及 AOP 的实现原理。在此基础上通过示例动手操作加深理解。

2.3.1　认识横切和纵切

首先认识一下什么是横切、纵切，这就要利用生物方面的知识了。切面的方向是这样规定的：拿植物的茎举例，纵切就是沿长轴来切，横切即是垂直与纵切的切法。编程相对来说是比较抽象的，有时候我们通过身边的事物来将抽象的具体化，这样也能更容易理解。

利用百度搜索“横切”“纵切”时，首先搜出来的结果是剖宫产的结果，内容是这样描述的：“由于人体和血管、神经系统等都是纵向的走向，所以纵切更有利于皮肤和伤口的愈合，通常在 1 年到 2 年就可以彻底恢复。但纵切的缺点是伤口较大，会留下明显的疤痕，会影响美观。横切的方法出血较少，并且出现伤口感染的机率要低于纵切，所以安全性比纵切高一些，只是不利再次进行剖宫产手术。”这里的横切、纵切与编程中的还是挺相似的。

2.3.2　什么是 AOP

AOP，可以说是 OOP（Object-Oriented Programing，面向对象编程）的补充和完善。OOP 引入封装、继承和多态性等概念来建立一种对象层次结构，用以模拟公共行为的一个集合。当我们需要为分散的对象引入公共行为（日志、安全、事务）的时候，OOP 则显得无能为力。也就是说，OOP 允许你定义从上到下的关系，但并不适合定义从左到右的关系。例如，日志功能。日志代码往往水平地散布在所有对象层次中，而与它所散布到的对象的核心功能毫无关系。对于其他类型的代码，如安全性、异常处理和透明的持续性也是如此。这种散布在各处的无关的代码被称为横切（cross-cutting）代码。在 OOP 设计中，它导致了大量代码重复、模块间的藕合度高，不利于各个模块的重用。

AOP 技术则恰恰相反，它利用一种称为“横切”的技术，剖解开封装的对象内部，并将那些影响了多个类的公共行为封装到一个可重用模块，并将其命名为“Aspect”，即切面。所谓“切面”，简单地说，就是将那些与业务无关，却为业务模块所共同调用的逻辑或责任封装起来，便于减少系统的重复代码，降低模块间的耦合度，并有利于未来的可操作性和可维护性。

2.3.3 AOP 原理

AOP 实际上是由目标类的代理类实现的。AOP 代理其实是由 AOP 框架动态生成的一个对象，该对象可作为目标对象使用。AOP 代理包含了目标对象的全部方法（见图 2-3），但 AOP 代理中的方法与目标对象的方法存在差异，AOP 方法在特定切入点添加了增强处理，并回调了目标对象的方法。

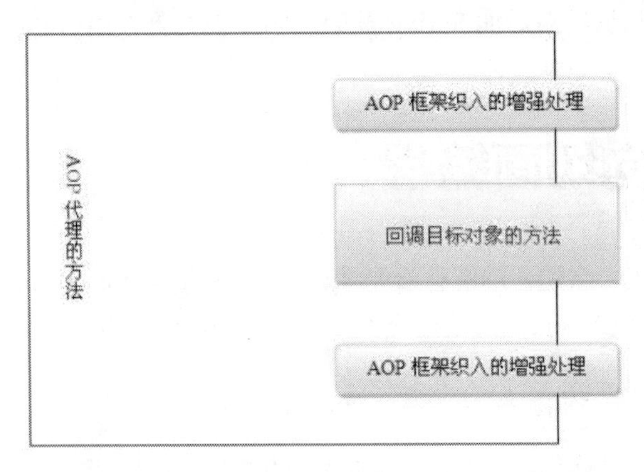

图 2-3

由于是代理实现 AOP，因此有必要学习一下代理。下面通过实例一步一步地了解静态代理和动态代理。新建一个 ServiceImplA 类，实现 IService 接口，想在调用 service 方法前后增加日志打印或为 service 方法增加 try catch，那么该怎么做呢？

```
package AOP;

public interface IService {
   public void service(String name) throws Exception;
}
package AOP;

public class ServiceImplA implements IService {

   @Override
   public void service(String name) throws Exception {
       System.out.println("ServiceImplA name"+name);
   }
}
```

1. 在每处调用的地方增加日志和 try catch

这也是一种方法，但缺点是很明显的，就是每处都要更改，量也会很大，显然不可取。这里是每个点都要加，一个方法可能被调用多处，就要写多次。而且以后再进行修改时也不方便，每个地方都要修改。

2. 代理模式

代理模式又分为动态代理模式和静态代理模式。

（1）静态代理

静态代理关键是在代理对象和目标对象实现共同的接口，并且代理对象持有目标对象的引用。这里用类ProxyServiceA来实现IService接口，同时将实现IService接口的对象作为一个属性。

```
package AOP;

public class ProxyServiceA implements IService {

    public ProxyServiceA(IService service) {
        super();
        this.service = service;
    }
    private IService service;

    public void service(String name) throws Exception {
        System.out.println("log start");
        try{
            service.service(name);
        }
        catch(Exception e)
        {
             throw e;
        }
        System.out.println("log end");
    }

   public static void main(String[] args) throws Exception {
        IService service=new ServiceImplA();
        service =new  ProxyServiceA(service);
        service.service("CYW");
   }
}
```

输出结果：

```
log start
ServiceImplA service:CYW
log end
```

有了ProxyServiceA之后，打印日志和增加try-catch只需放在ProxyServiceA类里面，便于后续修改，比如现在打印日志是输出在操作台的，哪天需要输入到日志文件时也只需修改ProxyServiceA中的打印操作即可。但问题来了：项目中的接口可不止一个，可能会有很多，而且每个接口中的方法也会有好多，这样一个一个地增加也是问题，于是有了动态代理。

（2）动态代理

在 Java 的动态代理机制中，有两个重要的类或接口：一个是 InvocationHandler(Interface)，另一个是 Proxy(Class)。这一个类和接口是实现动态代理所必须用到的。

```
package AOP;
import java.lang.reflect.InvocationHandler;
import java.lang.reflect.Method;
import java.lang.reflect.Proxy;
public class DynaProxyServiceA implements InvocationHandler {
    private Object target;//目标对象
    public Object bind(Object object){
        this.target = object;
        //生成动态代理对象
        Object obj=Proxy.newProxyInstance(this.target.getClass().
getClassLoader(), this.target.getClass().getInterfaces(), this);
        return obj;
    }
    public Object invoke(Object proxy, Method method, Object[] args) throws
Throwable {
        Object result = null;
        System.out.println("method:"+method);
        System.out.println("args:"+args);
        System.out.println("target:"+this.target);
        System.out.println("log start");
        try{
            result = method.invoke(this.target, args);
        }
        catch(Exception e)
        {
            throw e;
        }
        System.out.println("log end");
        return result;
    }
    public static void main(String[] args) throws Exception {
        IService service = (IService) new DynaProxyServiceA().bind(new
ServiceImplA());
        service.service("CYW");
    }
}
```

输出结果：

```
method:public abstract void AOP.IService.service(java.lang.String) throws java.lang.Exception
args:[Ljava.lang.Object;@1b6d3586
target:AOP.ServiceImplA@4554617c
log start
ServiceImplA service:CYW
log end

   method:public abstract void AOP.IService.service(java.lang.String) throws
java.lang.Exception
   args:[Ljava.lang.Object;@1b6d3586
   target:AOP.ServiceImplA@4554617c
   log start
   ServiceImplA service:CYW
   log end
```

通过 Proxy.newProxyInstance()生成的动态代理对象 A 都会与实现 InvocationHandler 接口的对象 B 关联，动态代理对象 A 调用目标对象方法时都会变成调用 B 中的 invoke 方法。在 invoke 方法中织入增强处理，并通过反射回调目标对象方法。在本例中，通过 bind()生成目标对象 ServiceImplA 的动态代理对象 A，A 关联了实现 InvocationHandler 接口对象的 DynaProxyServiceA，当动态代理对象 A 调用目标对象方法时会执行 DynaProxyServiceA 的 invoke 方法，增加 try-catch、打印日志，并回调目标对象的方法。

与前面的静态代理比较发现，动态代理不用再为每个接口手动创建代理类，其他对象只要与 InvocationHandler 接口对象 bind，就能获得该 InvocationHandler 接口对象的织入增强。

2.4 小结

我们回顾一下这一章节的主要内容，主要了解了 Spring 框架，学习了 IOC、AOP。在 Spring 框架部分，了解了框架的来源、七大模块。在 IOC 部分，介绍了依赖注入、控制反转以及两者的区别，以人与空气为例，使用 Spring 框架演示什么是依赖注入，也对 Spring 有一个大概的认识。在 AOP 部分，主要介绍了横切和纵切的概念、AOP 的概念，并通过实例来了解 AOP 的原理。

第 3 章

◀ 核心容器 ▶

在上一章节中，我们大致了解了 Spring 框架，并学习了控制反转（IOC）和面向切面编程（AOP）两个重要的编程思想，这一章我们将开始学习 Spring 框架中的核心容器。

本章主要涉及的知识点：

- IOC 容器：容器概念、BeanFactory、ApplicationContext。
- beans 的配置：三种配置方式。
- bean 的注入：三种注入方式以及常见数据类型注入介绍。
- bean 的生命周期和五大作用域。

为了能深入理解 Spring 框架，本章主要使用 XML 的方式来说明 Spring 框架的使用，虽然现在提倡使用注解方式配置，简单一些，但是为了更好、更深入地理解，这里还是建议要学习一下 XML 的配置方式。

3.1 IOC 容器

在第 2 章中介绍什么是依赖注入时有提到：依赖注入是组件之间的依赖关系由容器在运行期决定的，即由容器动态地将某个依赖关系注入组件之中。那什么是容器呢？既然 Spring 框架实现了 IOC，那么 Spring 中的容器是什么呢?

3.1.1 容器介绍

在日常生活中，容器是指用以容纳物料并以壳体为主的基本装置，是用来盛放东西的。在编程中，容器是用来存储和组织其他对象的对象，首先要确定容器也是对象，也可以当作 bean，只是这个对象是用来存储和组织其他对象的，那其他对象是什么呢?其他对象其实就是 bean 对象，这也是面向对象编程的一种体现，万物皆对象。Spring 提供了 BeanFactory、ApplicationContext 两个 IOC 容器来管理众多的 bean 对象。

3.1.2 BeanFactory

一提到工厂，我们可能就会想到富某康。工厂是一类用以生产货物的大型工业建筑物。

BeanFactory 不是用来生产货物的，而是用来生产管理 bean 的。BeanFactory 会在 bean 生命周期的各个阶段中对 bean 进行管理，并且 Spring 将这些阶段通过各种接口暴露给我们，让我们可以对 bean 进行各种处理。我们只要让 bean 实现对应的接口，那么 Spring 就会在 bean 的生命周期调用我们实现的接口来处理该 bean。这是怎么实现的呢？主要分为以下两个阶段。

1. bean 容器的启动

工厂要生产货物，首先得把工厂运转起来。同样，bean 容器要管理 bean，也需要先把容器启动起来，获取到 bean 的定义信息之后才能管理。

（1）读取 bean 的 xml 配置文件，然后将 xml 中每个 bean 元素分别转换成 BeanDefinition 对象。

```
    public abstract class AbstractBeanDefinition extends
BeanMetadataAttributeAccessor
    implements BeanDefinition, Cloneable {
    private volatile Object beanClass;
    private String scope = SCOPE_DEFAULT;
    private boolean abstractFlag = false;
    private boolean lazyInit = false;
    private int autowireMode = AUTOWIRE_NO;
    private int dependencyCheck = DEPENDENCY_CHECK_NONE;
    private String[] dependsOn;
    private ConstructorArgumentValues constructorArgumentValues;
    private MutablePropertyValues propertyValues;
    private String factoryBeanName;
    private String factoryMethodName;
    private String initMethodName;
    private String destroyMethodName;
```

BeanClass 保存 bean 的 class 属性，scope 保存 Bean 的作用域，abstractFlag 保存该 bean 是否抽象，lazyInit 保存是否延迟初始化，autowireMode 保存是否自动装配，dependencyCheck 保存是否坚持依赖，dependsOn 保存该 bean 依赖于哪些 bean（这些 bean 必须提取初始化），constructorArgumentValues 保存通过构造函数注入的依赖，propertyValues 保存通过 setter 方法注入的依赖，factoryBeanName 和 factoryMethodName 用于 factorybean，也就是工厂类型的 bean，initMethodName 和 destroyMethodName 分别对应 bean 的 init-method 和 destroy-method 属性。后面会对这些内容进行详细介绍。

（2）通过 BeanDefinitionRegistry 将 bean 注册到 beanFactory 中。

上面获取到 bean 的信息之后，是怎么注册到 BeanFactory 中的呢？其实是通过 BeanDefinitionRegistry 将 bean 注册到 beanFactory 中的，因为 BeanFactory 的实现类需要实现 BeanDefinitionRegistry 接口。

```
    public interface BeanDefinitionRegistry extends AliasRegistry {
       void registerBeanDefinition(String beanName, BeanDefinition
beanDefinition)throws BeanDefinitionStoreException;
```

```
        void removeBeanDefinition(String beanName) throws
    NoSuchBeanDefinitionException;
        BeanDefinition getBeanDefinition(String beanName) throws
    NoSuchBeanDefinitionException;
        boolean containsBeanDefinition(String beanName);
        String[] getBeanDefinitionNames();
        int getBeanDefinitionCount();
        boolean isBeanNameInUse(String beanName);
    }
```

BeanDefinitionRegistry 接口提供了根据 beanName 注册对应 beanDefinition 的方法。在 DefaultListableBeanFactory 类中实现了该方法，并将 beanDefinition 保存在了 ConcurrentHash-Map 中。

```
    @SuppressWarnings("serial")
    public class DefaultListableBeanFactory extends
AbstractAutowireCapableBeanFactory
            implements ConfigurableListableBeanFactory, BeanDefinitionRegistry,
Serializable {
        /** Map of bean definition objects, keyed by bean name */
        private final Map<String, BeanDefinition> beanDefinitionMap = new
ConcurrentHashMap<String, BeanDefinition>(64);

        @Override
        public void registerBeanDefinition(String beanName, BeanDefinition
beanDefinition)
                throws BeanDefinitionStoreException {
            // ... ...
            this.beanDefinitionMap.put(beanName, beanDefinition);
        }
```

另外，Spring 还对外暴露了一些接口来对 bean 初始化，例如 BeanFactoryPostProcessor。

```
    public interface BeanFactoryPostProcessor {
        /**
         * Modify the application context's internal bean factory after its
standard
         * initialization. All bean definitions will have been loaded, but no beans
         * will have been instantiated yet. This allows for overriding or adding
         * properties even to eager-initializing beans.
         * @param beanFactory the bean factory used by the application context
         * @throws org.springframework.beans.BeansException in case of errors
         */
        void postProcessBeanFactory(ConfigurableListableBeanFactory
beanFactory) throws BeansException;
    }
```

我们可以翻译一下 postProcessBeanFactory 的注释信息。postProcessBeanFactory 可以修改应用上下文中已经进行 standard 初始化的 beanFactory，此时所有 bean 的定义信息已经加载完成，但还未实例化，允许覆盖、新增甚至重新初始化 bean 信息，一个典型的例子就是属性覆盖器 PropertyOverrideConfigurer。对于一些参数，可以配置在 properties 中，而不用配置在 Spring 的 XML 配置文件中。

2. bean 容器的实例化

上面把 bean 容器启动之后，工厂算是运转起来了，配方（beanDefinition）也已经准备充分，然后就是生产（实例化）、管理货物（bean）了。实例化 bean 主要通过反射和 CGLIB 两种方式，在 bean 的实例化过程中，Spring 也暴露了一些接口，如表 3-1 所示。

表 3-1　在 bean 的实例化过程中 Spring 暴露的一些接口

接　口	说　明
BeanNameAware	获取该bean在配置文件中对应的id
BeanFactoryAware	获取实例化该bean的BeanFactory
InitializingBean	bean实例化、所有属性设置后调用初始化方法
DisposableBean	在bean丢弃的时候调用销毁方法

我们可以通过示例演示一下这几个接口的使用。

（1）创建 Maven project，在 pom.xml 中引入 spring-core、spring-context。

```
<project xmlns="http://maven.apache.org/POM/4.0.0"
xmlns:xsi="http://www.w3.org/2001/XMLSchema-instance"
    xsi:schemaLocation="http://maven.apache.org/POM/4.0.0
http://maven.apache.org/xsd/maven-4.0.0.xsd">
    <modelVersion>4.0.0</modelVersion>

    <groupId>com.demo</groupId>
    <artifactId>BeanFactoryDemo</artifactId>
    <version>0.0.1-SNAPSHOT</version>
    <packaging>jar</packaging>

    <name>BeanFactoryDemo</name>
    <url>http://maven.apache.org</url>
    <properties>
      <project.build.sourceEncoding>UTF-8</project.build.sourceEncoding>
      <spring.version>5.0.0.RELEASE</spring.version>
    </properties>
    <dependencies>
       <dependency>
          <groupId>org.springframework</groupId>
          <artifactId>spring-core</artifactId>
          <version>${spring.version}</version>
```

```
        </dependency>
        <dependency>
            <groupId>org.springframework</groupId>
            <artifactId>spring-context</artifactId>
            <version>${spring.version}</version>
        </dependency>
        <dependency>
          <groupId>junit</groupId>
          <artifactId>junit</artifactId>
          <version>3.8.1</version>
          <scope>test</scope>
        </dependency>
      </dependencies>
    </project>
```

（2）创建 bean 对象，实现上面列出的接口。

```
package com.demo.model;

import org.springframework.beans.BeansException;
import org.springframework.beans.factory.BeanFactory;
import org.springframework.beans.factory.BeanFactoryAware;
import org.springframework.beans.factory.BeanNameAware;
import org.springframework.beans.factory.DisposableBean;
import org.springframework.beans.factory.InitializingBean;

public class UserBean implements BeanNameAware,BeanFactoryAware,
InitializingBean,DisposableBean {

    public void setBeanName(String name) {
        System.out.println(name);
    }

    public void setBeanFactory(BeanFactory beanFactory) throws
BeansException {
        System.out.println(beanFactory);
    }

    public void afterPropertiesSet() throws Exception {
        System.out.println("InitializingBean");
    }

    public void destroy() throws Exception {
        System.out.println("DisposableBean");
    }
}
```

（3）bean 配置。

```
<?xml version="1.0" encoding="UTF-8"?>
<beans xmlns="http://www.springframework.org/schema/beans"
      xmlns:xsi="http://www.w3.org/2001/XMLSchema-instance"
      xsi:schemaLocation="http://www.springframework.org/schema/beans
http://www.springframework.org/schema/beans/spring-beans.xsd">
   <bean id="user" class="com.demo.model.UserBean"></bean>
</beans>
```

（4）测试。

使用 ApplicationContext 获取 BeanFactory，再通过 getBean 方法获取对应的 bean，最后调用 destroy 方法进行销毁，从输出结果可以看到依次调用了 BeanNameAware、BeanFactoryAware、InitializingBean、DisposableBean 接口。

```
public static void main( String[] args ) throws Exception
{
   ApplicationContext context=new ClassPathXmlApplicationContext(new
String[]{"applicationContext.xml"});
   BeanFactory factory=context;
   UserBean user=(UserBean)factory.getBean("user");
   user.destroy();
 }
```

输出结果：

```
user
org.springframework.beans.factory.support.DefaultListableBeanFactory@6bf
256fa: defining beans [user]; root of factory hierarchy
InitializingBean
DisposableBean
```


```
user
org.springframework.beans.factory.support.DefaultListableBeanFactory@6bf256fa: defining beans [user]; root of factory hierarchy
InitializingBean
DisposableBean
```


3.1.3 ApplicationContext

在上面的示例中，使用 ApplicationContext 获取了 bean 的配置，然后直接将 ApplicationContext 接口对象赋值给了 BeanFactory 接口对象，为什么可以赋值呢?其实 ApplicationContext 接口实现了 BeanFactory 接口。

```
public interface ApplicationContext extends EnvironmentCapable,
ListableBeanFactory, HierarchicalBeanFactory,MessageSource,
ApplicationEventPublisher, ResourcePatternResolver
```

从上面 ApplicationContext 接口的继承关系可以看到，它还通过继承其他接口扩展了 BeanFactory 的功能。

- MessageSource：为应用提供国际化访问功能。

- ResourcePattern Resolver：提供资源（如 URL 和文件系统）的访问支持。
- ApplicationEventPublisher：引入事件机制，包括启动事件、关闭事件等，让容器在上下文中提供了对应用事件的支持。它代表的是一个应用的上下文环境。

BeanFactory 主要是面对 Spring 框架的基础设施，面对 Spring 自己。Applicationcontext 主要面对于 Spring 使用的开发者。开发者基本都会使用 Applicationcontext，而非 BeanFactory，所以在上面实例中使用 ApplicationContext 获取 BeanFactory 接口对象。

上面的 ApplicationContext 对象是通过 ClassPathXmlApplicationContext 方法获取的，还有一种获取方式，即使用 FileSystemXmlApplicationContext 方法获取。

- ClassPathXmlApplicationContext 表示从类路径下加载配置文件。文件路径默认指的是项目的 classpath 路径下，所以不需要写前缀“classpath:”，如果指向绝对路径，需要加上“file:”。

```
new ClassPathXmlApplicationContext(new String[]
{"ApplicationContext.xml"});
```

- FileSystemXmlApplicationContext 表示从文件系统中加载配置文件。文件路径默认指的是项目的根目录下，若想使用项目的 classpath 路径，则需要加上 “classpath:” 。

```
new FileSystemXmlApplicationContext(new String[]{"classpath:
ApplicationContext.xml"});
```

3.2 Bean 的配置

在上一小节中为了演示容器 Bean 实例化时暴露出的几个接口，将 UserBean 配置在 XML 中，其实常见的 Bean 配置有 3 种：基于 XML 配置 Bean，使用注解定义 Bean，基于 Java 类提供 Bean 定义。

3.2.1 基于 XML 配置 Bean

基于 XML 配置 Bean 时，Spring 通过<bean>配置来实例化、设置 bean 的属性以及设置 bean 间的相互依赖性。一个<bean>通常需要定义 id 和 class 属性。class 属性是必需的，不然 Spring 怎么知道是哪个呢。id 不是必需的，不过如果配置文件中配置的其他 bean 要引用该 bean，那么 id 也是必需的，通常都加上。

```
<bean id="user" class="com.demo.model.UserBean"></bean>
```

3.2.2 使用注解定义 Bean

如果采用基于 XML 的配置，bean 定义信息和 bean 实现类本身是分离的，而采用基于注解的配置方式时，bean 定义信息即通过在 bean 实现类上标注注解实现。在第 2 章中，定义 Air 和 Person 时都使用了@Component 来注解 bean。

```
package com.demo.model;

import org.springframework.stereotype.Component;

@Component
public class CleanAir implements IAir {
    @Override
    public String toString() {

        return  "CleanAir";
    }
}
```

我们使用@Component 注解在 CleanAir 类声明处对类进行标注，它可以被 Spring 容器识别，自动将 POJO 转换为容器管理的 bean。它和<bean id="CleanAir" class="com.demo.model.CleanAir"></bean>是等效的。除了@Component 以外，Spring 提供了以下 3 个功能基本和@Component 等效的注解，分别用于对 DAO、Service 及 Web 层的 Controller 进行注解，所以也称这些注解为 Bean 的衍型注解：

- @Repository：用于对 DAO 实现类进行标注。
- @Service：用于对 Service 实现类进行标注。
- @Controller：用于对 Controller 实现类进行标注。

那问题来了：既然都是 bean，为什么还提供 4 个呢？因为这样能让注解的用途更加清晰，而且不同的注解也有各自特殊的功能。

3.2.3 基于 Java 类提供 Bean 定义

在普通的 POJO 类中只要标注@Configuration 注解，就可以为 Spring 容器提供 bean 定义的信息了。每个标注了@Bean 的类方法都相当于提供了一个 bean 的定义信息。

```
package com.demo.model;

import org.springframework.context.annotation.Bean;
import org.springframework.context.annotation.Configuration;

@Configuration
public class AppConf {

    @Bean
    public CleanAir cleanAir(){
       return new CleanAir();
    }
    @Bean
    public DirtyAir dirtyAir(){
       return new DirtyAir();
    }
```

```
    @Bean
    public Person person(){
        return new Person(dirtyAir());
    }
}
```

我们只需在 XML 中配置开启 context 扫描即可实现 bean 的配置，省去了配置 Person、DirtyAir。

```
<context:component-scan base-package="com.demo.model"/>
<bean id="conf" class="com.demo.model.AppConf"/>
    ApplicationContext context=new ClassPathXmlApplicationContext(new
String[]{"ApplicationContext.xml"});
    BeanFactory factory=context;
    AppConf conf=(AppConf)factory.getBean("conf");
    conf.person().Breath();
```

输出结果：

```
DirtyAir
```

基于 Java 类的配置方式和基于 XML 或基于注解的配置方式相比，前者通过代码的方式更加灵活地实现了 Bean 的实例化及 Bean 之间的装配，后面两者是通过配置声明的方式实现的，在灵活性上要稍逊一些，但是配置上要更简单一些。

3.3 Bean 的注入

在 Bean 的配置中介绍的是 Bean 的声明问题，即在哪声明、怎么声明的问题。本节将要讲解的 Bean 的注入是怎么实例化、怎么注入的问题。Bean 注入的方式有两种：一种是在 XML 中配置，另一种是使用注解的方式注入。

3.3.1 XML 方式注入

XML 方式注入一般有三种方式：属性注入、构造函数注入和工厂方法注入。

1. 属性注入

在传统的对象实例化时可以先使用 new class(),然后通过 setXXX()方法设置对象的属性值或依赖对象，属性注入也是采用这种方式，只是 Spring 框架会在内部完成这些操作，它会先调用 Bean 的默认构造函数实例化 Bean 对象,然后通过反射的方式调用 Setter 方法注入属性值。它会使用默认的构造函数（无参数构造函数），只需为注入的属性设置 set 方法，可选择性和灵活性比较高，所以也是比较常用的一种注入方式。这里还是在 IOC 章节人和空气的基础上稍作修改来演示。IAir 接口和 CleanAir、DirtyAir 类不变，这里就不贴了。

（1）新建 XMLInstance 类

```
package com.demo.model;

public class XMLInstance{

    private String name;

    public void setName(String name) {
       this.name = name;
    }
   private IAir air;

   public void setAir(IAir air) {
      this.air = air;
   }
    public void Breath()
    {
       System.out.println("Name:"+this.name+";Air:"+this.air.toString());
    }
}
```

在 XMLInstance 类中并未声明构造函数，对于 air 属性只设置了 set 方法，get 方法也没设置。

```
<bean id="CleanAir" class="com.demo.model.CleanAir">
    <qualifier value="cleanair"/>
</bean>
<bean id="xmlinstance" class="com.demo.model.XMLInstance">
    <property name="air" ref="CleanAir"></property>
    <property name="name" value="abc"></property>
</bean>
```

在 XML 中使用 property 类配置属性，name 表示属性名，value 用来设置基本数据类型的属性值。在 Spring 配置文件中，bean 之间可以相互引用，引用时可以用<ref>标签配置 bean 的 id 属性。<ref>可以用在<property>属性中，也可以用在<construct-arg>构造函数的参数值中，还可以用在其他地方，通过引用能减少 bean 的声明。

2. 构造函数注入

在属性注入时先使用默认的构造函数（无参数构造函数）实例化，然后通过 set 方法注入属性，在传统实例化对象时可以自定义构造函数进行实例化。构造函数注入就是通过自定义构造函数来进行对象的实例化。这里在 XMLInstance 类的基础上增加了一个构造函数，第一个参数是 String 类型的 name，第二个参数是 IAir 类型的 air。

```
public XMLInstance(String name, IAir air) {
    super();
    this.name = name;
```

```
        this.air = air;
    }
```

在 XML 中，使用<construect-arg>来设置构造函数的参数，index 属性设置参数的顺序，参数顺序应该与构造函数一致，ref 设置引用 bean 的 id，value 设置构造函数参数的值。

```
<bean id="xmlcontructinstance" class="com.demo.model.XMLInstance">
    <constructor-arg index="1" ref="CleanAir"></constructor-arg>
    <constructor-arg index="0" value="abc"></constructor-arg>
</bean>
```

3. 工厂方法注入

工厂方法注入参考的是工厂设计模式，在工厂类中实现对象的实例化。工厂类负责创建一个或多个目标类实例，一般以接口或抽象类变量的形式返回目标类实例。工厂类对外屏蔽了目标类的实例化步骤，调用者甚至不用知道具体的目标类是什么。工厂方法也分静态工厂方法和非静态工厂方法。静态工厂方法不用实例化工厂类，直接通过类名调用。非静态工厂方法需要先实例化工厂类，然后通过工厂类对象调用获取对象。这里创建了一个工厂类 XMLFactory，在类中定义了一个静态方法和一个实例方法（用来实例化 bean 对象）。

```
package com.demo.model;

public class XMLFactory {

    public XMLInstance CreateInstance()
    {
        return new XMLInstance("instance",new CleanAir());
    }

    public static XMLInstance CreateStaticInstance()
    {
        return new XMLInstance("static instance",new CleanAir());
    }
}
```

（1）静态工厂方法

只需设置工厂方法对应的类，以及对应的工厂方法。

```
<bean id="xmlfactorystaticinstance" class="com.demo.model.XMLFactory"
factory-method="CreateStaticInstance"></bean>
```

（2）实例工厂方法

需要先实例化工厂类，再通过工厂类对象调用实例方法获取 bean 对象。

```
<bean id="xmlfactory" class="com.demo.model.XMLFactory"></bean>
<bean id="xmlfactoryinstance" factory-bean="xmlfactory"
factory-method="CreateInstance"></bean>
```

4. 常见数据类型注入

（1）List 属性注入

使用<list>配置 java.util.List 类型的属性。List 属性中的元素可以是任何数据类型的值，如果是 Java 对象，可以使用 ref 指定，或使用<bean>定义新实例。如果是基础数据类型，可直接使用字符串。<list>中的元素会按配置的先后顺序排序。

```
<property name="lists">
     <list>
         <value>1</value>
         <ref bean="CleanAir" />
         <bean class="com.demo.model.CleanAir"/>
    </list>
</property>
```

（2）Set 属性注入

使用<set>配置 java.util.Set 类型的属性。Set 属性中的元素可以是任何数据类型的值，如果是 Java 对象，可以使用 ref 指定，或使用<bean>定义新实例。如果是基础数据类型，可直接使用字符串。<set>中的元素没有先后顺序。

```
</property>
   <property name="sets">
    <set>
       <value>1</value>
       <ref bean="CleanAir" />
        <bean class="com.demo.model.CleanAir"/>
    </set>
</property>
```

（3）Map 属性注入

使用<map>配置 java.util.Map 类型的属性。<entry>配置 Map 里的元素，key 指定索引，value 指定值。如果是 Java 对象，可以使用 ref 指定，或使用<bean>定义新实例。

```
<property name="maps">
    <map>
     <entry key="key1" value="1"></entry>
        <entry key="key2" value-ref="CleanAir"></entry>
        <entry key="key3" >
             <bean class="com.demo.model.CleanAir"/>
        </entry>
    </map>
</property>
```

（4）Properties 属性注入

使用<props>配置 java.util.Properties 类型的属性。<props>配置一个 Properties 对象，<prop>配置一条属性，属性 key 配置索引。

```
<property name="pros">
    <props>
        <prop key="prokey1">prokeyA</prop>
        <prop key="prokey2">prokeyB</prop>
    </props>
</property>
```

（5）自定义属性编辑器

有一些属性是没法注入的，此时就需要自定义，比如日期类型。可以通过继承 PropertyEditorSupport 的类，重写 setAsText 方法来实现注入。这里定义了 CustomerProperty 继承 PropertyEditorSupport，重写了 setAsText 方法，并将该 bean 配置到 XML 中。

```
package com.demo.model;

import java.beans.PropertyEditorSupport;
import java.text.ParseException;
import java.text.SimpleDateFormat;

public class CustomerProperty extends PropertyEditorSupport {
    private String format="yyyy-MM-dd";
    public String getFormat() {
        return format;
    }
    public void setFormat(String format) {
        this.format = format;
    }
    @Override
    public void setAsText(String text) throws IllegalArgumentException {

        SimpleDateFormatsdf=new SimpleDateFormat(format);
        //super.setAsText(text);
        try {
            //转换对象，通过 setValue 方法重新赋值
            this.setValue(sdf.parse(text));
        } catch (ParseException e) {
            e.printStackTrace();
        }
    }
}
<bean id="customEditorConfigurer" class="org.springframework.beans.
factory.config.CustomEditorConfigurer">
    <property name="customEditors">
        <map>
            <entry key="java.util.Date" value="com.demo.model.
CustomerProperty"/>
        </map>
    </property>
```

```
</bean>
```

配置之后就可以注入 Date 类型的属性了。

```
<property name="date" value="2018-8-20"/>
```

新建 XmlCollectionsDemo 类，配置上面几个类型的属性来演示。

```
package com.demo.model;

import java.util.Date;
import java.util.List;
import java.util.Map;
import java.util.Properties;
import java.util.Set;

import org.springframework.beans.factory.BeanFactory;
import org.springframework.context.ApplicationContext;
import
org.springframework.context.support.ClassPathXmlApplicationContext;

public class XmlCollectionsDemo {

    private List<Object> list;

    private Properties pros;

    private Set<Object> sets;

    private Map<String,Object> maps;

    private Date date;

    public Date getDate() {
        return date;
    }

    public void setDate(Date date) {
        this.date = date;
    }

    public List<Object> list() {
        return list;
    }

    public void setLists(List<Object> list) {
        this.list = list;
    }
```

```
    public Properties getPros() {
        return pros;
    }

    public void setPros(Properties pros) {
        this.pros = pros;
    }

    public Set<Object> getSets() {
        return sets;
    }

    public void setSets(Set<Object> sets) {
        this.sets = sets;
    }

    public Map<String, Object> getMaps() {
        return maps;
    }

    public void setMaps(Map<String, Object> maps) {
        this.maps = maps;
    }

    public static void main( String[] args ) throws Exception
     {

        ApplicationContext context=new ClassPathXmlApplicationContext(new
String[]{"ApplicationContext.xml"});
        BeanFactory factory=context;
        XmlCollectionsDemo annontationInstance=(XmlCollectionsDemo)
factory.getBean("xmlCollectionsDemo");
        System.out.println(annontationInstance.list);
        System.out.println(annontationInstance.pros);
        System.out.println(annontationInstance.sets);
        System.out.println(annontationInstance.maps);
        System.out.println(annontationInstance.date);
     }
  }
```

下面是属性对应的 XML 配置文件。

```
<bean id="xmlCollectionsDemo" class="com.demo.model.XmlCollectionsDemo">
<property name="lists">
    <list>
        <value>1</value>
```

```
            <ref bean="CleanAir" />
             <bean class="com.demo.model.CleanAir"/>
        </list>
    </property>
       <property name="sets">
         <set>
            <value>1</value>
            <ref bean="CleanAir" />
             <bean class="com.demo.model.CleanAir"/>
        </set>
    </property>
    <property name="maps">
        <map>
         <entry key="key1" value="1"></entry>
            <entry key="key2" value-ref="CleanAir"></entry>
            <entry key="key3" >
                 <bean class="com.demo.model.CleanAir"/>
            </entry>
        </map>
    </property>
    <property name="pros">
        <props>
            <prop key="prokey1">prokeyA</prop>
            <prop key="prokey2">prokeyB</prop>
        </props>
    </property>
    <property name="date" value="2018-8-20"/>
  </bean>
```

通过运行 main 方法，打印出属性值：

```
[1, CleanAir, CleanAir]
{prokey2=prokeyB, prokey1=prokeyA}
[1, CleanAir, CleanAir]
{key1=1, key2=CleanAir, key3=CleanAir}
Mon Aug 20 00:00:00 CST 2018
```

5. 初始化函数、销毁函数

通过上面几种注入方式的学习，对通过 XML 对 bean 实例化有了了解。有的对象在实例化之后还需要执行某些初始化代码，但这些初始化代码还不能写在构造函数中，此时可以将初始化代码写到某个方法中。将 init-method 属性值设置为该方法，Spring 会强制执行该方法进行初始化。有的对象在使用完毕之后需要释放，可以使用 destroy-method 来进行销毁。

```
public void DestoryMethod()
{
    System.out.println("DestoryMethod");
}
```

```
public void InitMethod()
{
    System.out.println("InitMethod");
}
```

这里先在 XMLInstance 类中增加上面两个方法来模拟销毁和初始化方法，然后在 xml 配置 bean 时设置 destroy-method、init-method 属性值对应两个方法的方法名。注解中的@PostConstruct 对应 init-method、@PreDestory 对应 destroy-method。

```
<bean id="xmlfactoryinstance" factory-bean="xmlfactory" factory-method=
"CreateInstance" destroy-method="DestoryMethod" init-method="InitMethod">
</bean>
```

3.3.2 注解方式注入

1. 常用注解介绍

学习完 XML 注入之后再学习注解方式注入就容易得多了，注解方式注入主要涉及@Autowired、@Resource、@Required、@Qualifier、@Value 这几个注解。在 2.2.4 节定义 Person 时就使用过@Autowired、@Qualifier。下面来了解一下它们的具体用法。

- @Autowired：默认按类型匹配注入 bean，可以对类成员变量、方法及构造函数进行标注，完成自动装配的工作。在使用@Autowired 时，首先在容器中查询对应类型的 bean。如果查询结果刚好为一个，就将该 bean 装配给@Autowired 指定的数据；如果查询的结果不止一个，那么@Autowired 会根据名称来查找。如果查询的结果为空，那么会抛出异常。解决方法时，使用 required=false。
- @Required：适用于 bean 属性 setter 方法，并表示受影响的 bean 属性必须在 XML 配置文件在配置时进行填充；否则，容器会抛出一个 BeanInitializationException 异常。
- @Qualifier：@Autowired 默认是单实例的，但是在面向接口编程中，如果把一个属性设置为接口类型，一个接口就可能有多个实现，那么到底注入哪一个呢?为了解决这个问题，就有了@Qualifier。
- @Value：在 xml 配置属性时可以通过 property 的 value 设置默认值，@Value 也可以为属性设置默认值。
- @Resource：默认按名称匹配注入 bean。要求提供一个 bean 名称的属性，如果属性为空，就自动采用标注处的变量名或方法名作为 bean 的名称。如果我们没有在使用@Resource 时指定 bean 的名字，同时 Spring 容器中又没有该名字的 bean，这时@Resource 就会退化为@Autowired，即按照类型注入。

```
@Component
public class AnnontationInstance {

    @Value("abc")
    private String name;


    public void setName(String name) {
```

```
        this.name = name;
    }

    //@Resource与@Autowired两者选其一
    // @Autowired
    // @Qualifier(value="cleanair")
    private IAir air;
    @Resource(name="CleanAir")
    public void setAir(IAir air) {
        this.air = air;
    }

    public void Breath()
    {
        System.out.println("Name:"+this.name+";Air:"+this.air.toString());
    }
}
```

在上面的代码中，使用@Value 注解为 name 设置了默认值，使用@Resources 设置 bean 的名称为 IAir 属性注入 bean，也可以使用@Autowired+@Qualifier 为 IAir 注入 bean。

2. 开启注解

配置完注解之后，还要告诉 Spring 开启注解，这样@Autowired、@Resources 这些注解才起作用。开启注解有两种比较简单的方式。

（1）在 xml 配置文件中使用 context:annotation-config：

```
<context:annotation-config />
```

（2）在 xml 配置文件中使用 context:component-scan：

```
<context:component-scan base-package="com.demo.model"/>
```

3.4 Bean 的作用域和生命周期

在 3.1.2 节容器启动获得 BeanDefinition 对象中有一个 scope 属性。该属性控制着 bean 对象的作用域。本节介绍 Bean 的作用域及生命周期，了解 bean 是怎么来的、怎么没的。

3.4.1 Bean 的作用域

在 3.1.2 节中介绍到 Bean 容器启动会先读取 bean 的 xml 配置文件，然后将 xml 中每个 bean 元素分别转换成 BeanDefinition 对象。在 BeanDefinition 对象中有 scope 属性，就是它控制着 bean 的作用域。

Spring 框架支持 5 种作用域，有 3 种作用域是当开发者使用基于 Web 的 ApplicationContext 时才生效的。表 3-2 给出了 Spring 直接支持的作用域。当然开发者也可以自己定制作用域。

表 3-2　Spring 直接支持的作用域

作　用　域	描　　述
单例（singleton）	默认，每一个 Spring IoC 容器都拥有唯一的一个实例对象
原型（prototype）	一个 Bean 定义，任意多个对象
请求（request）	一个 HTTP 请求会产生一个 Bean 对象，也就是说，每一个 HTTP 请求都有自己的 Bean 实例，只在基于 Web 的 Spring ApplicationContext 中可用
会话（session）	限定一个 Bean 的作用域为 HTTPSession 的生命周期。同样，只有基于 Web 的 Spring ApplicationContext 才能使用
全局会话（global session）	限定一个 Bean 的作用域为全局 HTTPSession 的生命周期。通常用于门户网站场景，同样，只有基于 Web 的 Spring ApplicationContext 可用

我们可以以 XMLInstance 类为基础演示一下 singleton 和 prototype 作用域。

这里使用通过 BeanFactory 的 getBean 方法获取两次 bean 对象。

```
XMLInstance instance=(XMLInstance)factory.getBean("xmlinstance");
instance.setName("123");
instance.Breath();
instance=(XMLInstance)factory.getBean("xmlinstance");
instance.Breath();
```

如果我们采用 bean 默认的作用域 singleton，如下配置，那么两个 getBean 获取的对象是一致的。

```
    <bean id="xmlinstance" class="com.demo.model.XMLInstance"
scope="singleton">
      <property name="air" ref="CleanAir"></property>
      <property name="name" value="abc"></property>
    </bean>
```

输出结果：

```
Name:123;Air:CleanAir
Name:123;Air:CleanAir
```

如果我们采用 bean 默认的作用域 prototype，如下配置，那么两个 getBean 获取的对象是不一致的。

```
    <bean id="xmlinstance" class="com.demo.model.XMLInstance"
scope="prototype">
      <property name="air" ref="CleanAir"></property>
      <property name="name" value="abc"></property>
    </bean>
```

输出结果：

```
Name:123;Air:CleanAir
Name:abc;Air:CleanAir
```

3.4.2　Bean 的生命周期

前面章节介绍了 bean 容器以及 bean 的配置与注入，本节学习 bean 的生命周期（见图 3-1），了解 bean 是怎么来的、怎么没的。

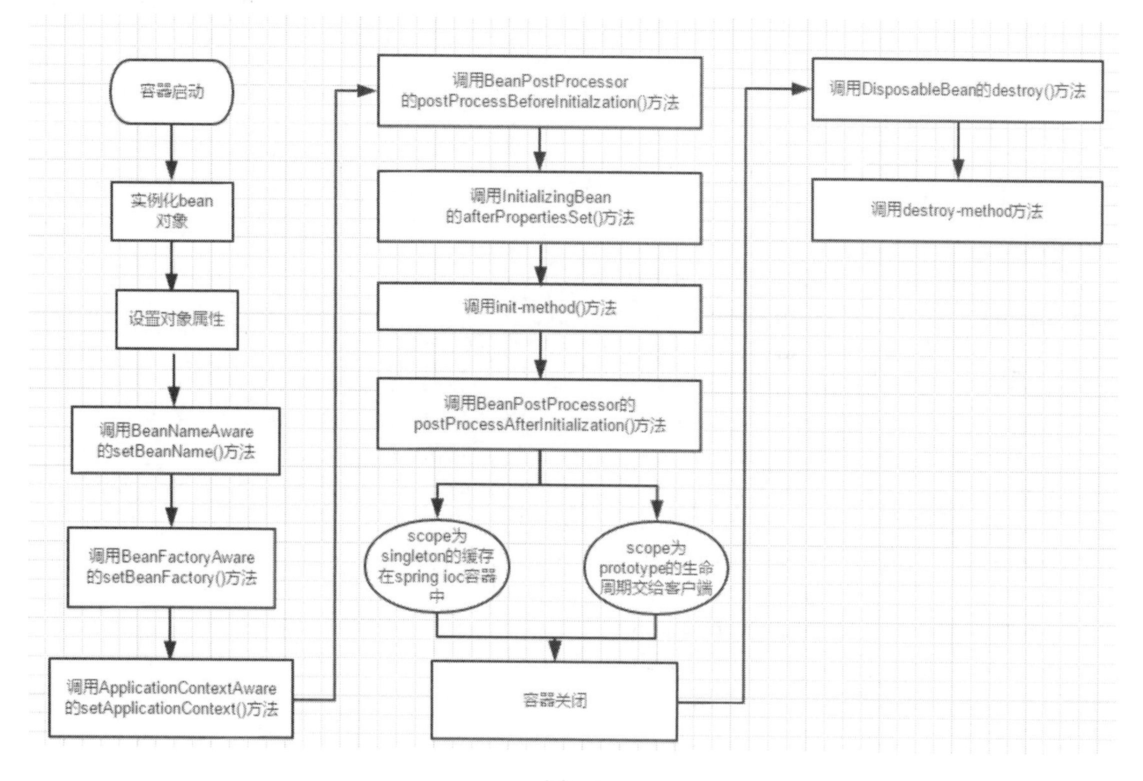

图 3-1

在 ApplicationContext 容器中，Bean 的生命周期流程如图 3-1 所示。

（1）容器启动后，会对 scope 为 singleton 且非懒加载的 bean 进行实例化。

（2）按照 Bean 定义信息配置信息，注入所有的属性。

（3）如果 Bean 实现了 BeanNameAware 接口，会回调该接口的 setBeanName()方法，传入该 Bean 的 id，此时该 Bean 就获得了自己在配置文件中的 id。

（4）如果 Bean 实现了 BeanFactoryAware 接口，会回调该接口的 setBeanFactory()方法，传入该 Bean 的 BeanFactory，这样该 Bean 就获得了自己所在的 BeanFactory。

（5）如果 Bean 实现了 ApplicationContextAware 接口，会回调该接口的 setApplicationContext()方法，传入该 Bean 的 ApplicationContext，这样该 Bean 就获得了自己所在的 ApplicationContext。

（6）如果有 Bean 实现了 BeanPostProcessor 接口，就会回调该接口的 postProcessBefore-Initialzation()方法。

（7）如果 Bean 实现了 InitializingBean 接口，就会回调该接口的 afterPropertiesSet()方法。

（8）如果 Bean 配置了 init-method 方法，就会执行 init-method 配置的方法。

（9）如果Bean实现了BeanPostProcessor接口，就会回调该接口的postProcessAfterInitialization()方法。

（10）之后，就可以正式使用该 Bean 了。对于 scope 为 singleton 的 Bean，Spring 的 ioc 容器中会缓存一份该 Bean 的实例；对于 scope 为 prototype 的 Bean，每次被调用都会新建一个新的对象，其生命周期就交给调用方管理了，不再是由 Spring 容器进行管理了。

（11）容器关闭后，如果 Bean 实现了 DisposableBean 接口，就会回调该接口的 destroy()方法。

（12）如果 Bean 配置了 destroy-method 方法，就会执行 destroy-method 配置的方法。至此，整个 Bean 的生命周期结束。

这里在 UserBean 类基础上进行改造，增加 name 属性并实现 ApplicationContextAware 接口。

```
package com.demo.model;

import org.springframework.beans.BeansException;
import org.springframework.beans.factory.BeanFactory;
import org.springframework.beans.factory.BeanFactoryAware;
import org.springframework.beans.factory.BeanNameAware;
import org.springframework.beans.factory.DisposableBean;
import org.springframework.beans.factory.InitializingBean;
import org.springframework.context.ApplicationContext;
import org.springframework.context.ApplicationContextAware;

public class UserBean implements BeanNameAware,BeanFactoryAware,
InitializingBean,DisposableBean,ApplicationContextAware {

    private String name;

    public String getName() {
        return name;
    }

    public void setName(String name) {
        this.name = name;
        System.out.println("set 方法被调用");
    }

    public UserBean() {
         System.out.println("UserBean 类构造方法");
    }

    public void setBeanName(String name) {
        System.out.println("BeanNameAware 被调用");
```

```
        }

        public void setBeanFactory(BeanFactory beanFactory) throws
BeansException {
            System.out.println("BeanFactoryAware 被调用");
        }

        public void afterPropertiesSet() throws Exception {
            System.out.println("InitializingBean 被调用");
        }

        public void destroy() throws Exception {
            System.out.println("DisposableBean 被调用");
        }
        //自定义的初始化函数
        public void myInit() {
            System.out.println("myInit 被调用");
        }
        //自定义的销毁方法
        public void myDestroy() {
            System.out.println("myDestroy 被调用");
        }

       public void setApplicationContext(ApplicationContext applicationContext)
throws BeansException {
            System.out.println("setApplicationContext 被调用");
        }

    }
```

定义后置处理器CusBeanPostProcessor，实现BeanPostProcessor接口。

```
    package com.demo.model;

    import org.springframework.beans.BeansException;
    import org.springframework.beans.factory.config.BeanPostProcessor;

    public class CusBeanPostProcessor implements BeanPostProcessor {

        public Object postProcessBeforeInitialization(Object bean, String
beanName) throws BeansException {
            System.out.println("postProcessBeforeInitialization 被调用");
            return bean;

        }

        public Object postProcessAfterInitialization(Object bean, String
beanName) throws BeansException {
```

```
            System.out.println("postProcessAfterInitialization 被调用");
            return bean;
        }

    }
```

在 xml 中配置 bean 和 BeanPostProcessor。

```
    <bean id="user" class="com.demo.model.UserBean" destroy-method="myDestroy"
init-method="myInit">
        <property name="name" value="abc"></property>
    </bean>
    <bean id="postProcessor" class="com.demo.model.CusBeanPostProcessor" />
```

测试：

```
    ApplicationContext context=new ClassPathXmlApplicationContext(new
String[]{"applicationContext.xml"});
    BeanFactory factory=context;
    UserBean user=(UserBean)context.getBean("user");
    ((ClassPathXmlApplicationContext)context).close();
```

输出结果如图 3-2 所示。

```
UserBean 类构造方法
set 方法被调用
BeanNameAware 被调用
BeanFactoryAware 被调用
setApplicationContext 被调用
postProcessBeforeInitialization 被调用
InitializingBean 被调用
myInit 被调用
postProcessAfterInitialization 被调用
DisposableBean 被调用
myDestroy 被调用
```

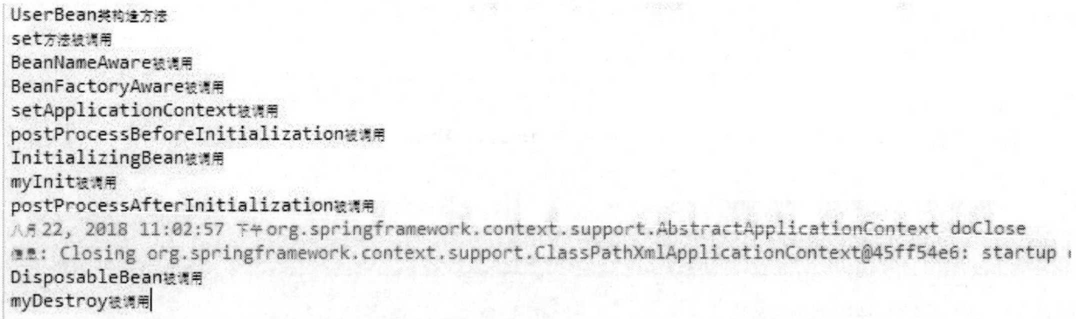

```
UserBean类构造方法
set方法被调用
BeanNameAware被调用
BeanFactoryAware被调用
setApplicationContext被调用
postProcessBeforeInitialization被调用
InitializingBean被调用
myInit被调用
postProcessAfterInitialization被调用
八月 22, 2018 11:02:57 下午 org.springframework.context.support.AbstractApplicationContext doClose
信息: Closing org.springframework.context.support.ClassPathXmlApplicationContext@45ff54e6: startup
DisposableBean被调用
myDestroy被调用
```

图 3-2

3.5 小结

本章主要介绍了 Bean 容器 BeanFactory 和 ApplicationContext，学习了 Bean 的配置和注入、Bean 的作用域和生命周期。通过本章的学习，了解了 Spring 最核心、最基本的模块，为后面的学习打下基础。

第 4 章

◂ Spring之AOP ▸

在上一章节中，我们大致了解了 Spring 核心容器，了解了 IOC 思想在 Spring 中的具体应用 Bean 容器以及 Bean 的配置与使用，这一章我们将开始学习 AOP 在 Spring 框架中的具体应用。

本章主要涉及的知识点：

- AOP 的引入：从传统处理方法到 AOP 处理，一步一步引入。
- AOP 的实现方式：通过实践了解 AOP 的 3 种实现方式。

4.1 AOP 基础

在第 2 章中也有介绍 AOP，不过侧重点是 AOP 的原理和动态代理，本节主要介绍 AOP 的基础知识点。

4.1.1 AOP 的引入

这里可以把单个模块当作一个圆柱，假如没有 AOP，那么在做日志处理的时候，我们就会在每个模块中添加日志或者权限处理，日志或权限类似圆柱体的部分圆柱，如图 4-1 所示。

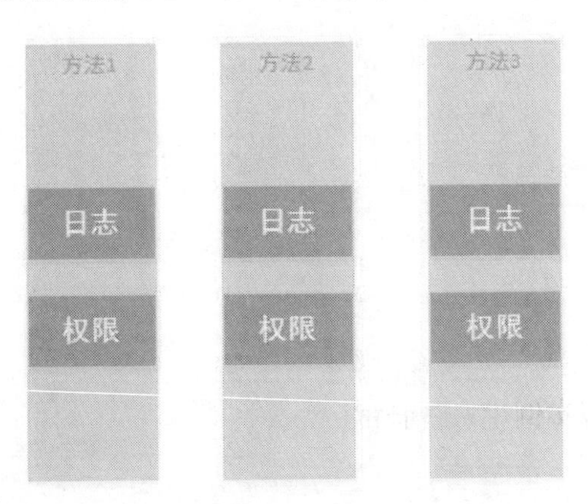

图 4-1

一般大多数的日志或权限处理代码是相同的，为了实现代码复用，我们可能把日志处理抽离成一个新的方法，如图 4-2 所示。

即使这样，我们仍然必须手动插入这些方法，而且这两个方法是强耦合的。假如此时我们不需要这个功能了，或者想换成其他功能，就必须一个个修改。

通过动态代理，可以在指定位置执行对应流程。这样就可以将一些横向的功能抽离出来，形成一个独立的模块，然后在指定位置插入这些功能。这样的思想被称为面向切面编程，亦即 AOP，如图 4-3 所示。

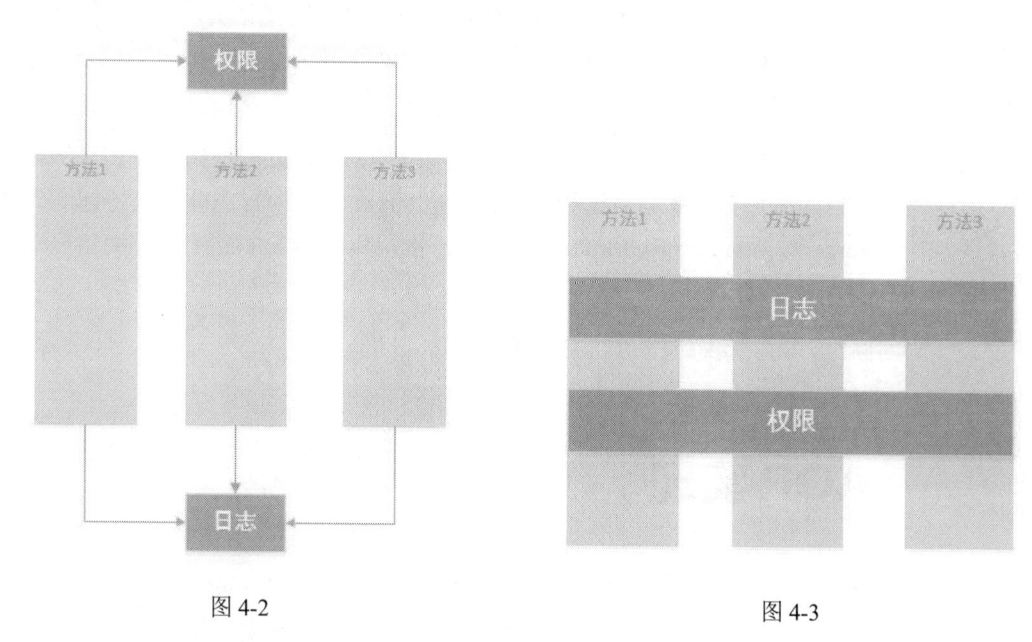

图 4-2　　　　图 4-3

4.1.2　AOP 主要概念

4.1.1 小节介绍了引入 AOP 的好处，本小节来了解一下 AOP 的几个核心概念。

（1）横切关注点

AOP 把一个业务流程分成几部分，例如权限检查、业务处理、日志记录，每个部分单独处理，然后把它们组装成完整的业务流，每部分被称为切面或关注点。

（2）切面

类是对物体特征的抽象，切面就是对横切关注点的抽象。可以将每部分抽象成一叠纸一样，一层一层的，那么每张纸都是一个切面。

（3）连接点

因为 Spring 只支持方法类型的连接点，所以在 Spring 中连接点指的就是被拦截到的方法，实际上连接点还可以是字段或者构造器。其实，Spring 只支持方法类型的连接点包含字段和构造器。因为字段通过 get、set 方法得到，构造器其实也是方法。Spring 只支持方法类型的连接点和连接点是字段或者构造器是包含关系。

（4）切入点

对连接点进行拦截的定义。连接点可以有很多，但并不一定每个连接点都进行操作，比如莲藕，藕段与藕段之间是有连接点的，但不一定都切开。

（5）通知

通知指的就是指拦截到连接点之后要执行的代码，分为前置、后置、异常、最终、环绕通知 5 类。这个有点类似于把藕段与藕段断开之后要做的事情，是往里面加蜂蜜还是做什么。

（6）目标对象

代理的目标对象，就是动态代理的 target，在实际操作中一般会先实现 AOP 的接口，然后配置这些接口作用到哪些对象上，被作用的对象就是目标对象。

（7）织入

切面是独立的，目标对象也是独立的，它们是不耦合的，那它怎么把切面放到目标对象中呢？这时就需要进行织入操作，就类似这种的，怎么把 target 和打印日志联系到一起呢？使用动态代理。在 Spring 中，aop.framework.ProxyFactory 用作织入器，进行横切逻辑的织入。

（8）引入

不改代码的同时，为类动态地添加方法或字段。

4.2 AOP 实现

4.1 节主要介绍了 AOP 的理论知识，本节通过示例进一步理解 Spring 中 AOP 的使用，主要介绍 AOP 的 3 种实现方式：经典的基于代理的 AOP、AspectJ 基于 XML 的配置、AspectJ 基于注解的配置。

1. 经典的基于代理的 AOP

基于代理的 AOP 主要介绍 MethodBeforeAdvice、AfterReturningAdvice、ThrowsAdvice 三个接口的使用。

- MethodBeforeAdvice：见名知意，通过方法名就可以猜到它的作用。方法前拦截器在执行指定方法前调用，参数分别为被调用的方法、执行时被传入的参数、被拦截的 Bean。
- AfterReturningAdvice：返回后拦截器，在执行完指定方法并返回之后调用。如果有返回值可以获取到返回值，否则为 null。参数分别为方法返回值、被调用的方法、执行时被传入的参数、被拦截的 Bean。
- ThrowsAdvice：异常拦截器，在指定方法抛出异常时被调用。该接口并未定义方法，因此不需要实现任何方法。那它是怎么拦截的呢？我们可以查看该接口的定义，在定义类文档中有如图 4-4 所示的说明。例如，在实现该接口的类中定义了 public void afterThrowing(Exception ex)、public void afterThrowing(Method method, Object[] args, Object target, Exception ex)方法抛出异常时就会被调用。

```
19⊖/**
20  * Tag interface for throws advice.
21  *
22  * <p>There are not any methods on this interface, as methods are invoked by
23  * reflection. Implementing classes must implement methods of the form:
24  *
25  * <pre class="code">void afterThrowing([Method, args, target], ThrowableSubclass);</pre>
26  *
27  * <p>Some examples of valid methods would be:
28  *
29  * <pre class="code">public void afterThrowing(Exception ex)</pre>
30  * <pre class="code">public void afterThrowing(RemoteException)</pre>
31  * <pre class="code">public void afterThrowing(Method method, Object[] args, Object target, Exception ex)</pre>
32  * <pre class="code">public void afterThrowing(Method method, Object[] args, Object target, ServletException ex)</pre>
33  *
34  * The first three arguments are optional, and only useful if we want further
35  * information about the joinpoint, as in AspectJ <b>after-throwing</b> advice.
36  *
37  * <p><b>Note:</b> If a throws-advice method throws an exception itself, it will
38  * override the original exception (i.e. change the exception thrown to the user).
39  * The overriding exception will typically be a RuntimeException; this is compatible
40  * with any method signature. However, if a throws-advice method throws a checked
41  * exception, it will have to match the declared exceptions of the target method
42  * and is hence to some degree coupled to specific target method signatures.
43  * <b>Do not throw an undeclared checked exception that is incompatible with
44  * the target method's signature!</b>
```

图 4-4

在软件开发中推荐面向接口的编程，所以这里创建一个 IAOPServices 接口，并定义两个方法：withAopMethod 方法将使用拦截器拦截的方法，withNoAopMethod 方法不会被拦截器拦截。接口代码如下：

```
package basicAop;

public interface IAOPServices {

   public String withAopMethod() throws Exception;

   public String withNoAopMethod() throws Exception;

}
```

在 AOPServicesImpl 类中实现了该接口，并在该类中定义了 String 类型的 description 属性，以及对应的 getter、setter 方法。两个接口方法将返回该 description 属性的值。

```
package basicAop;

public class AOPServicesImpl implements IAOPServices {

   private String description;
   public String getDescription() {
       return description;
   }

   public void setDescription(String description) {
       this.description = description;
   }
```

```
    public String withAopMethod() throws Exception {
        System.out.println("AOP 函数运行方法:withAopMethod");
        if(description.trim().length()==0){
            throw new Exception("description 属性不能为空");
        }
        return description;
    }

    public String withNoAopMethod() throws Exception {
        System.out.println("无 AOP 函数运行方法:withNoAopMethod");
        return description;
    }
}
```

上面把要使用 AOP 拦截的方法准备好了，下面就是定义 AOP 拦截器方法了。这里在 AOPInterceptor 类中实现了上面的 AfterReturningAdvice、MethodBeforeAdvice、ThrowsAdvice 三个接口。

```
package basicAop;

import java.lang.reflect.Method;

import org.springframework.aop.AfterReturningAdvice;
import org.springframework.aop.MethodBeforeAdvice;
import org.springframework.aop.ThrowsAdvice;

public class AOPInterceptor implements
AfterReturningAdvice,MethodBeforeAdvice,ThrowsAdvice {

    public void afterReturning(Object value, Method method, Object[] args,
Object instance) throws Throwable {

        System.out.println("方法"+method.getName()+"运行结束,返回值为:"
+value);
    }

    public void before(Method method, Object[] args, Object instance) throws
Throwable {
        System.out.println("执行 MethodBeforeAdvice，即将执行的方法:"
+method.getName());
        if(instance instanceof AOPServicesImpl)
        {
            String
description=((AOPServicesImpl)instance).getDescription();
            if(description==null)
            {
                throw new NullPointerException("description 属性不能为 null");
```

```
            }
        }
    }
    public void afterThrowing(Exception ex){
        System.out.println("抛出了异常:"+ex.getMessage());
    }

    public void afterThrowing(Method method, Object[] args, Object target,
Exception ex){
            System.out.println("方法"+method.getName()+"抛出了异常:"
+ex.getMessage());
    }
}
```

这里要拦截的方法和拦截器都准备好了，怎么将拦截器用于拦截该方法呢？这里就需要进行配置了。首先在 pom.xml 中引入依赖，这里引入 spring-aop、spring-context。

```
<project xmlns="http://maven.apache.org/POM/4.0.0"
xmlns:xsi="http://www.w3.org/2001/XMLSchema-instance"
xsi:schemaLocation="http://maven.apache.org/POM/4.0.0
http://maven.apache.org/xsd/maven-4.0.0.xsd">
    <modelVersion>4.0.0</modelVersion>
    <groupId>com.demo</groupId>
    <artifactId>BasicAOP</artifactId>
    <version>0.0.1-SNAPSHOT</version>
    <properties>
      <project.build.sourceEncoding>UTF-8</project.build.sourceEncoding>
      <spring.version>5.0.0.RELEASE</spring.version>
    </properties>
  <dependencies>
      <dependency>
          <groupId>org.springframework</groupId>
          <artifactId>spring-context</artifactId>
          <version>${spring.version}</version>
      </dependency>
     <dependency>
         <groupId>org.springframework</groupId>
         <artifactId>spring-aop</artifactId>
         <version>${spring.version}</version>
     </dependency>
  </dependencies>
  </project>
```

实际上 Spring 无法将 Services 实现类与拦截器直接组装，因为没有对应的 setter、getter 方法。安装时，先借助 Spring 中的代理类将自定义拦截器注入 NameMatchMethodPointcutAdvisor 类的 advice 属性中，再将定义好的 NameMatchMethodPointcutAdvisor 对象注入 ProxyFactoryBean 中。这里将自定义的 AOPInterceptor 拦截器注入 NameMatchMethodPointcutAdvisor 中，然后将

NameMatchMethodPointcutAdvisor 对象注入 ProxyFactoryBean 中。

```
<?xml version="1.0" encoding="UTF-8"?>
<beans xmlns="http://www.springframework.org/schema/beans"
     xmlns:xsi="http://www.w3.org/2001/XMLSchema-instance"
     xmlns:context="http://www.springframework.org/schema/context"
     xmlns:aop="http://www.springframework.org/schema/aop"
     xsi:schemaLocation="http://www.springframework.org/schema/beans
        http://www.springframework.org/schema/beans/spring-beans.xsd
        http://www.springframework.org/schema/context
        http://www.springframework.org/schema/context/spring-
context.xsd
        http://www.springframework.org/schema/aop
        http://www.springframework.org/schema/aop/spring-aop.xsd">

  <bean id="aopInterceptor" class="org.springframework.aop.support.
NameMatchMethodPointcutAdvisor">
    <property name="advice">
        <bean class="basicAop.AOPInterceptor"></bean>
    </property>
    <property name="mappedName" value="withAopMethod"></property>
  </bean>

  <bean id="aopService" class="org.springframework.aop.framework.
ProxyFactoryBean">
    <property name="interceptorNames">
        <list>
        <value>aopInterceptor</value>
        </list>
    </property>
    <property name="target">
        <bean class="basicAop.AOPServicesImpl">
            <property name="description" value="basicAop"></property>
        </bean>
    </property>
  </bean>
</beans>
```

从 Spring 容器中获取 IAOPServices 对象，并分别执行 IAOPServices 中的两个方法。Spring 会在 withAopMethod()方法前后添加拦截器，在 withNoAopMethod()方法前后并不会添加。

```
public class AopTest {

   public static void main(String[] args) throws Exception {

       ApplicationContext context=new ClassPathXmlApplicationContext(new
String[]{"applicationContext.xml"});
       BeanFactory factory=context;
```

```
        IAOPServices services=(IAOPServices)context.getBean("aopService");
        services.withAopMethod();
        services.withNoAopMethod();
    }
}
```

运行结果如下：

```
执行MethodBeforeAdvice,即将执行的方法:withAopMethod
AOP函数运行方法:withAopMethod
方法withAopMethod运行结束,返回值为:basicAop
无AOP函数运行方法:withNoAopMethod
```

2. AspectJ 基于 XML 的配置

AspectJ 是一个面向切面的框架，扩展了 Java 语言。AspectJ 定义了 AOP 语法，有一个专门的编译器，用来生成遵守 Java 字节编码规范的 Class 文件。我们还是利用 IAOPServices 接口、AOPServicesImpl 类实现 AspectJ 基于 XML 的 AOP 编程。表 4-1 给出 AspectJ 主要的配置元素。使用 AspectJ 时需要引入两个依赖项，即 aopalliance、aspectjweaver。在引入这两个依赖项时需要注意，有时会报错，更新两个依赖项的版本就好了。

```
<dependency>
    <groupId>aopalliance</groupId>
    <artifactId>aopalliance</artifactId>
    <version>1.0</version>
</dependency>
<!-- https://mvnrepository.com/artifact/org.aspectj/aspectjweaver -->
<dependency>
    <groupId>org.aspectj</groupId>
    <artifactId>aspectjweaver</artifactId>
    <version>1.8.11</version>
</dependency>
```

表 4-1　AspectJ 主要的配置元素

AOP 配置元素	描　　述
<aop:config>	顶层的 AOP 配置元素，大多数的<aop:*>元素必须包含在<aop:config>元素内
<aop:aspect>	定义切面
<aop:aspect-autoproxy>	启用@AspectJ 注解驱动的切面
<aop:pointcut>	定义切点
<aop:advisor>	定义 AOP 通知器
<aop:before>	定义 AOP 前置通知
<aop:after>	定义 AOP 后置通知（不管被通知的方法是否执行成功）
<aop:after-returning>	定义成功返回后的通知

（续表）

AOP 配置元素	描　述
<aop:after-throwing>	定义抛出异常后的通知
<aop:around>	定义 AOP 环绕通知
<aop:declare-parents>	为被通知的对象引入额外的接口，并透明地实现

这里定义了 XMLAdvice 拦截器方法，用于演示前置、后置、成功返回、异常返回、环绕通知。

```
package Services;
import org.aspectj.lang.JoinPoint;
import org.aspectj.lang.ProceedingJoinPoint;

public class  XMLAdvice {
    public void beforeAdvice() {
        System.out.println("前置通知执行了");
    }

    public void afterAdvice() {
        System.out.println("后置通知执行了");
    }

    public void afterReturnAdvice(String result) {
        System.out.println("返回通知执行了" + "运行业务方法返回的结果为" +
result);
    }

    public String aroundAdvice(ProceedingJoinPoint proceedingJoinPoint)
throws Throwable {
        String result = "";
        try {
            System.out.println("环绕通知开始执行了");
            long start = System.currentTimeMillis();
            result = (String) proceedingJoinPoint.proceed();
            long end = System.currentTimeMillis();
            System.out.println("环绕通知执行结束了");
            System.out.println("执行业务方法共计: " + (end - start) + "毫秒。");
        } catch (Throwable e) {

        }
        return result;
    }

    public void throwingAdvice(JoinPoint joinPoint, Exception e) {
        StringBuffer stringBuffer = new StringBuffer();
        stringBuffer.append("异常通知执行了.");
```

```
            stringBuffer.append("方法:").append(joinPoint.getSignature().
getName()).append("出现了异常.");
            stringBuffer.append("异常信息为:").append(e.getMessage());
            System.out.println(stringBuffer.toString());
        }

    }
```

上面把拦截器定义完了，之后就是把定义好的拦截器与 Services 关联在一起。使用 AOP 配置元素将 Services 与拦截器中的方法关联上。

```
    <bean id="serviceImplA" class="Services.AOPServicesImpl">
       <property name="description" value="basicAop"></property>
    </bean>
    <bean id="serviceAspectBean" class="Services.XMLAdvice" />
    <aop:config>
        <aop:aspect id="serviceAspect" ref="serviceAspectBean">
            <aop:pointcut id="servicePointcut" expression="execution(*
Services.*.withAop*(..))" />
            <aop:before pointcut-ref="servicePointcut" method="beforeAdvice" />
            <aop:after pointcut-ref="servicePointcut" method="afterAdvice" />
            <aop:after-returning pointcut-ref="servicePointcut"
method="afterReturnAdvice" returning="result" />
            <aop:around pointcut-ref="servicePointcut" method="aroundAdvice" />
            <aop:after-throwing pointcut-ref="servicePointcut"
method="throwingAdvice" throwing="e" />
        </aop:aspect>
    </aop:config>
    </beans>
```

配置完之后还是和经典的基于代理的 AOP 一样，运行代码从 Spring 容器中获取 IAOPServices 对象，并分别执行 IAOPServices 中的两个方法。Spring 会在 withAopMethod()方法前后添加拦截器，在 withNoAopMethod()方法前后并不会添加。

```
    public class App {

       public static void main(String[] args) throws Exception {

            ApplicationContext context=new ClassPathXmlApplicationContext(new
String[]{"applicationContext.xml"});
            BeanFactory factory=context;
            IAOPServices services=(IAOPServices)context.getBean("serviceImplA");
            services.withAopMethod();
            services.withNoAopMethod();
        }
    }
```

运行结果如下：

```
信息: Overriding bean definition for bean 'se
前置通知执行了
环绕通知开始执行了
AOP函数运行方法:withAopMethod
环绕通知执行结束了
执行业务方法共计：0毫秒。
返回通知执行了运行业务方法返回的结果为basicAop
后置通知执行了
无AOP函数运行方法:withNoAopMethod
```

3. AspectJ 基于注解的配置

AspectJ 基于 XML 的配置还是需要在 XML 中配置 AOP 元素。现在一般提倡使用注解来进行编程，AspectJ 也提供了基于注解的实现方式。基于注解的 AOP 配置其实和基于 XML 的一样，可以参照基于 XML 的来进行理解。这里定义了 AnnontationAdvice，并用@Aspect 注解定义切面。在 XML 中的配置元素改成了注解关键字。

```
package Services;

import org.aspectj.lang.JoinPoint;
import org.aspectj.lang.ProceedingJoinPoint;
import org.aspectj.lang.annotation.After;
import org.aspectj.lang.annotation.AfterReturning;
import org.aspectj.lang.annotation.AfterThrowing;
import org.aspectj.lang.annotation.Around;
import org.aspectj.lang.annotation.Aspect;
import org.aspectj.lang.annotation.Before;
import org.springframework.stereotype.Component;

@Component
@Aspect
public class AnnontationAdvice {

   @Before("execution(* Services.*.withAop*(..))")
   public void beforeAdvice() {
       System.out.println("前置通知执行了");
    }

   @After("execution(* Services.*.withAop*(..))")
    public void afterAdvice() {
       System.out.println("后置通知执行了");
    }

   @AfterReturning(value="execution(* Services.*.withAop*(..))",
returning="result")
```

```
        public void afterReturnAdvice(String result) {
            System.out.println("返回通知执行了" + "运行业务方法返回的结果为" +
result);
        }

        @Around(value="execution(* Services.*.withAop*(..))")
        public String aroundAdvice(ProceedingJoinPoint proceedingJoinPoint)
throws Throwable {
            String result = "";
            try {
                System.out.println("环绕通知开始执行了");
                long start = System.currentTimeMillis();
                result = (String) proceedingJoinPoint.proceed();
                long end = System.currentTimeMillis();
                System.out.println("环绕通知执行结束了");
                System.out.println("执行业务方法共计: " + (end - start) + "毫秒。");
            } catch (Throwable e) {

            }
            return result;
        }
        @AfterThrowing(value="execution(* Services.*.withAop*(..))",
throwing="e")
        public void throwingAdvice(JoinPoint joinPoint, Exception e) {
            StringBuffer stringBuffer = new StringBuffer();
            stringBuffer.append("异常通知执行了.");
            stringBuffer.append("方法:").append(joinPoint.getSignature().
getName()).append("出现了异常.");
            stringBuffer.append("异常信息为:").append(e.getMessage());
            System.out.println(stringBuffer.toString());
        }
    }
```

在配置文件中只需配置一下自动扫描的包名，并配置<aop:aspectj-autoproxy>即可，比 XML 的配置简单一些。

```
<!-- 配置自动扫描的包 -->
<context:component-scan base-package="Services"></context:component-scan>
<!-- 自动为切面方法中匹配的方法所在的类生成代理对象。 -->
<aop:aspectj-autoproxy></aop:aspectj-autoproxy>
```

最后从 Spring 容器中获取 IAOPServices 对象，并分别执行 IAOPServices 中的两个方法。运行结果如下，从打印的日志可以看到拦截器拦截了 withAopMethod()方法，并未拦截 withNoAopMethod():

```
信息: Overriding bean definition for bean
环绕通知开始执行了
前置通知执行了
AOP函数运行方法:withAopMethod
环绕通知执行结束了
执行业务方法共计：1毫秒。
后置通知执行了
返回通知执行了运行业务方法返回的结果为basicAop
无AOP函数运行方法:withNoAopMethod
```

4.3 小结

本章主要从传统处理方式一步一步地引入 AOP，介绍 AOP 的主要概念，并列举了 AOP 的 3 种实现。通过本章的学习，对 AOP 思想有了更深入的认识。其实，实现 AOP 还有其他方式，这里就不一一列举了。

第 5 章

Spring之DAO

在上一章节中，我们了解了 Spring 框架中的 AOP 模块，这一章节我们开始学习 Spring 框架中的 DAO 模块。

本章主要涉及的知识点：

- JDBC 基本用法：Statement、PreparedStatement、CallableStatement 的使用。
- JDBC 高级用法：批处理、事务处理。
- Spring DAO 模块：JdbcDaoSupport、MappingSqlQuery 等对象的使用。
- Spring 事务管理：TransactionProxyFactoryBean 、DataSourceTransactionManager 的配置与使用。

5.1 JDBC 详解

在了解 Spring 的 DAO 模块时需要有一定的数据库基础，Java 语言与数据库连接使用的是 JDBC，所以有必要学习一下 JDBC 的内容。

5.1.1 JDBC 介绍

JDBC（Java DB Connection，Java 数据库连接）是一种可用于执行 SQL 语句的 Java API（Application Programming Interface，应用程序设计接口）。它由一些 Java 语言编写的类和界面组成。JDBC 为数据库应用开发人员和数据库前台工具开发人员提供了一种标准的应用程序设计接口，使开发人员可以用纯 Java 语言编写完整的数据库应用程序。JDBC 代表 Java 数据库连接。它是一个软件层，允许开发者在 Java 中编写客户端/服务器应用。

通过使用 JDBC，开发人员可以很方便地将 SQL 语句传送给几乎任何一种数据库。也就是说，开发人员可以不必写一个程序访问 Sybase，写另一个程序访问 Oracle，再写一个程序访问 Microsoft 的 SQL Server。用 JDBC 写的程序能够自动将 SQL 语句传送给相应的数据库管理系统（DBMS）。不但如此，使用 Java 编写的应用程序可以在任何支持 Java 的平台上运行，不必在不同的平台上编写不同的应用。Java 和 JDBC 的结合可以让开发人员在开发数据库应用时真正实现“Write Once，Run Everywhere！”。

5.1.2 操作步骤

JDBC 可以连接不同的数据库，不同的数据库也可以被不同的工具连接，但在连接时基本都是固定的几个步骤。

1. 驱动引入

JDBC 是对外开放的接口，数据库提供商实现了这些接口，这些接口的组合就是驱动。数据库有好多种，例如 MySQL、Oracle 等，需要注册不同的驱动来操作对应的数据库，注册驱动也得要有驱动才是，所以首先要将驱动引入项目。

2. 注册驱动

引入驱动之后应用程序也不知道是用的什么数据库，只是把驱动下载了下来放到项目中，所以得注册一下才知道是谁，注册之后会返回对应的驱动管理对象。就和入职一样，你到公司了但不报到，那也不知道来了没来，报到了才会有针对个人的流程。

3. 创建连接

数据库和应用程序是分隔开来的，数据库可能存放在远程，那怎么和数据库搭上呢？这就需要连接了。

4. 执行操作

连接上之后要干什么呢，不能一直连着不干事情啊，这也是资源的一种浪费，所以连接之后执行数据库的操作增、删、改、查等。

5. 返回结果

增、删、改、查操作结束之后，总要有一个结果，不然怎么知道成功与否，查询的话会返回查询的数据，增加、删除、修改会返回影响的行数。

6. 释放资源

把结果也返回了，但不能老连着数据库，这样占用资源，创建的对象也没有释放，还占空间，所以用完了就把它关掉。

5.1.3 Statement 的使用

Statement 是 Java 执行数据库操作的一个重要接口，用于在已经建立数据库连接的基础上，向数据库发送要执行的 SQL 语句。Statement 对象用于执行静态 SQL 语句，并返回它所生成结果的对象。

默认情况下，同一时间每个 Statement 对象只能打开一个 ResultSet 对象。因此，如果读取一个 ResultSet 对象与读取另一个交叉，那么这两个对象必须是由不同的 Statement 对象生成的。如果存在某个语句打开的当前 ResultSet 对象，那么 Statement 接口中的所有执行方法都会隐式关闭它。

在使用 Statement 之前先进行数据准备，这里在本地 MySQL 中创建了一个数据库 daodemodb 和一张表 t_user，并在表中增加了几条数据用来测试。

```
/* 数据库创建 */
CREATE DATABASE `daodemodb` /*!40100 DEFAULT CHARACTER SET utf8 */;
/*创建测试表*/
CREATE TABLE `t_user` (
  `id` int(11) NOT NULL AUTO_INCREMENT,
  `name` varchar(45) DEFAULT NULL,
  `age` int(11) DEFAULT NULL,
  `money` decimal(10,2) DEFAULT NULL,
  PRIMARY KEY (`id`)
) ENGINE=InnoDB DEFAULT CHARSET=utf8;
/*数据准备*/
insert into t_user(name,age,money)
values
('张三','24',666.66),
('李四','25',888.88),
('王二','26',999.99),
('小明','27',555.55),
('小赵','28',333.33)
```

按照上面的操作步骤，需要引入驱动，这里使用 pom.xml 引入 mysql 的 jdbc 驱动。

```
<dependency>
    <groupId>mysql</groupId>
    <artifactId>mysql-connector-java</artifactId>
    <version>8.0.12</version>
</dependency>
```

然后就是在代码中依次注册驱动、创建连接、执行操作、返回结果、释放资源步骤，下面代码演示的就是这个过程，从 t_user 表中查询数据并打印到日志中。

```
public static void main( String[] args ) throws SQLException
{
    Connection conn=null;
    Statement stmt=null;
    ResultSet rs=null;
    try{
     //注册驱动
     DriverManager.registerDriver(new com.mysql.cj.jdbc.Driver());
     //通过注册的驱动获得连接对象 Connection
  conn=DriverManager.getConnection("jdbc:mysql://127.0.0.1:3306/
daodemodb?
  useUnicode=true&characterEncoding=UTF-8"
              + "&serverTimezone=UTC&useSSL=false","root","123456");
     //通过 Statement 对象执行操作 返回结果 ResultSet
```

```
            stmt=conn.createStatement();
            //返回结果
            rs=stmt.executeQuery("select * from t_user");
            while(rs.next()){
                System.out.println("姓名:"+rs.getString("name")+"  年龄:"+
rs.getInt("age") +"  余额:"+rs.getDouble("money"));
                }
            }
            catch(SQLException e){
            System.out.println(e.getMessage());
            e.printStackTrace();
            }
            finally{
            //释放资源
            if(conn!=null){ conn.close(); }
            if(stmt!=null){ stmt.close(); }
               if(rs!=null){ rs.close(); }
               }
        }
```

输出结果：

```
姓名:张三  年龄:24  余额:666.66
姓名:李四  年龄:25  余额:888.88
姓名:王二  年龄:26  余额:999.99
姓名:小明  年龄:27  余额:555.55
姓名:小赵  年龄:28  余额:333.33
```

5.1.4 使用 PreparedStatement 返回自增主键

实际上有三种 Statement 对象，它们都作为在给定连接上执行 SQL 语句的包容器：Statement、PreparedStatement（从 Statement 继承而来）和 CallableStatement（从 PreparedStatement 继承而来）。它们都专用于发送特定类型的 SQL 语句：Statement 对象用于执行不带参数的简单 SQL 语句；PreparedStatement 对象用于执行带或不带 IN 参数的预编译 SQL 语句；CallableStatement 对象用于执行对数据库已存在的存储过程的调用。Statement 接口提供了执行语句和获取结果的基本方法。PreparedStatement 接口添加了处理 IN 参数的方法；而 CallableStatement 添加了处理 OUT 参数的方法。

这里向 t_user 表中插入一条数据，并返回自增主键 id 的值。在准备 SQL 时使用？来做参数的占位符，在实例化 PreparedStatement 对象之后对 SQL 进行传参，这样也能防止注入式攻击。

```
public static void main(String[] args) throws SQLException {
    // TODO Auto-generated method stub
    Connection conn=null;
```

```
        PreparedStatement prestmt=null;
        ResultSet rs=null;
        try{
            //注册驱动
            DriverManager.registerDriver(new com.mysql.cj.jdbc.Driver());
            //通过注册的驱动获得连接对象 Connection
            conn=DriverManager.getConnection("jdbc:mysql://127.0.0.1:3306/
daodemodb?useUnicode=true&characterEncoding=UTF-8"+ "&serverTimezone=
UTC&useSSL=false","root","123456");

            //PreparedStatement 对象
            String sql="insert into t_user (name,age,money) values(?,?,?)";
     prestmt=conn.prepareStatement(sql,Statement.RETURN_GENERATED_KEYS);
            prestmt.setString(1, "小李");
            prestmt.setInt(2, 25);
            prestmt.setDouble(3,222.22);
            //返回结果
            int result=prestmt.executeUpdate();
            if(result>0)
            {
                System.out.println("新增成功");
                rs=prestmt.getGeneratedKeys();
                while(rs.next())
                {
                   System.out.println("生成的主键 ID 为:"+rs.getInt(1));
                }
             }
        }
        catch(SQLException e){
            System.out.println(e.getMessage());
            e.printStackTrace();
        }
        finally{
            //释放资源
            if(conn!=null){ conn.close(); }
            if(prestmt!=null){ prestmt.close(); }
            if(rs!=null){ rs.close(); }
        }
    }
```

输出结果：

```
新增成功
生成的主键 ID 为:6
```

5.1.5 使用 CallableStatement 调用存储过程

在使用数据库的过程中，可能会调用存储过程，可以使用 CallableStatement 来调用存储过程。

- 调用存储函数：{?= call <procedure-name>[(<arg1>,<arg2>, ...)]}。
- 调用存储过程：{call <procedure-name>[(<arg1>,<arg2>, ...)]}。

通过 CallableStatement 对象的 registerOutParameter() 方法注册 Out 参数。通过 CallableStatement 对象的 setXxx()方法设定 IN 或 In out 参数，若想将参数设为 null，可以使用 setNull()。如果所调用的是带返回参数的存储过程，还需要通过 CallableStatement 对象的 getXxx()获取输出参数。

在数据库中创建了存储过程 p_selectUserByAge，根据用户年龄查找用户，存储过程一个传入参数 age，一个传出参数 count，参数 count 存放根据年龄查找的用户个数。

```
CREATE  PROCEDURE `p_selectUserByAge`(age int, out count int)
BEGIN
  set count=(select count(1) from t_user t where t.age =age);
  select * from  t_user t where t.age =age;
END
```

在下面的代码中先使用 Connection 的 prepareCall 方法来实例化 CallableStatement，再使用 CallableStatement 对象的 registerOutParameter 方法设置传入参数，最后执行存储过程返回结果。

```
public static void main(String[] args) throws SQLException {

    Connection conn=null;
    CallableStatement callstmt=null;
    ResultSet rs=null;
    try{
        //注册驱动
        DriverManager.registerDriver(new com.mysql.cj.jdbc.Driver());
        //通过注册的驱动获得连接对象 Connection
        conn=DriverManager.getConnection("jdbc:mysql://127.0.0.1:3306/
daodemodb?useUnicode=true&characterEncoding=UTF-8"
                + "&serverTimezone=UTC&useSSL=false","root","123456");
        //CallableStatement 对象
          callstmt=conn.prepareCall("{call p_selectUserByAge(?,?)}");
          callstmt.setInt(1,25);
          //设置传入参数
          callstmt.registerOutParameter(2, Types.INTEGER);
          rs=callstmt.executeQuery();
          while(rs.next())
          {
```

```
                System.out.println("姓名:"+rs.getString("name")+"  年龄:" +
rs.getInt("age") +"  出生日期:"+rs.getDouble("money"));
            }
            System.out.println("存储过程返回值:"+callstmt.getInt(2));
        }
        catch(SQLException e){
            System.out.println(e.getMessage());
            e.printStackTrace();
        }
        finally{
            //释放资源
            if(conn!=null){ conn.close(); }
            if(callstmt!=null){ callstmt.close(); }
          if(rs!=null){ rs.close(); }
        }
    }
```

输出结果：

```
姓名:李四  年龄:25  出生日期:888.88
姓名:小李  年龄:25  出生日期:222.22
存储过程返回值:2
```

5.1.6　批处理

在实际开发中往往会批量执行 SQL，Statement 和 PreparedStatement 都支持批量执行 SQL 语句，但这些 SQL 必须是 Insert、Update、Delete 这种执行后返回一个 Int 类型的数，表示影响的行数。Statement 和 PreparedStatement 都是通过 addBatch()方法添加一条 SQL 语句，通过 executeBatch()方法批量执行 SQL 语句，返回一个 Int 类型的数组，表示各 SQL 的返回值，这样就减少了注入驱动、创建连接这些步骤，提升了效率。首先看一下 Statement 批处理的例子：

```
public static void main(String[] args) throws SQLException {
    // TODO Auto-generated method stub
      Connection conn=null;

      ResultSet rs=null;
      Statement stmt=null;
      try{
        //注册驱动
        DriverManager.registerDriver(new com.mysql.jdbc.Driver());
        //通过注册的驱动获得连接对象 Connection
        conn=DriverManager.getConnection("jdbc:mysql://127.0.0.1:3306/
daodemodb?useUnicode=true&characterEncoding=UTF-8"
              + "&serverTimezone=UTC&useSSL=false","root","123456");
```

```
            stmt=conn.createStatement();
            for(int i=0;i<2;i++)
            {
                String sql="insert into t_user (name,age,money)
values('StatementTest"+i+"',"+25+i+",222.22)";
                stmt.addBatch(sql);
            }
            //批处理
            int [] result=stmt.executeBatch();
            System.out.println("影响的行数分别为:");
            for(int i=0;i<result.length;i++)
            {
                System.out.print(result[i]+"  ");
            }
        }
        catch(SQLException e)
        {
            System.out.println(e.getMessage());
            e.printStackTrace();
        }
        finally
        {
            //释放资源
            if(conn!=null)
            {
                conn.close();
            }
            if(stmt!=null)
            {
                stmt.close();
            }
            if(rs!=null)
            {
                rs.close();
            }
        }
    }
```

由于 Statement 无法传递参数，必须是完整的 SQL 语句，因此先将 SQL 拼接之后通过 addBatch(sql)方法加入到批处理中，然后通过 executeBatch 方法执行批处理返回影响行数的数组。

输出结果：

```
影响的行数分别为:
1  1
```

PreparedStatement 既可以是完整的 SQL，也可以用带参数的不完整的 SQL。我们看一下使用 PreparedStatement 进行批处理的例子。

```
public static void main( String[] args ) throws SQLException
{
    Connection conn=null;

    ResultSet rs=null;
    PreparedStatement prestmt=null;
    try{
        //注册驱动
        DriverManager.registerDriver(new com.mysql.jdbc.Driver());
        //通过注册的驱动获得连接对象 Connection
        conn=DriverManager.getConnection("jdbc:mysql://127.0.0.1:3306/
daodemodb?useUnicode=true&characterEncoding=UTF-8"
            + "&serverTimezone=UTC&useSSL=false","root","123456");
        String sql="insert into t_user (name,age,money) values(?,?,?)";
        prestmt=conn.prepareStatement(sql);
        for(int i=0;i<2;i++)
        {
            prestmt.setString(1, "PreparedStatementTest"+i);
            prestmt.setInt(2, 25+i);
            prestmt.setDouble(3,222.22);
            prestmt.addBatch();
        }
        //批处理
        int [] result=prestmt.executeBatch();
        System.out.println("影响的行数分别为:");
        for(int i=0;i<result.length;i++)
        {
            System.out.print(result[i]+"  ");
        }
    }
    catch(SQLException e)
    {
        System.out.println(e.getMessage());
        e.printStackTrace();
    }
    finally
    {
        //释放资源
        if(conn!=null)
        {
            conn.close();
```

```
            }
            if(prestmt!=null)
            {
                prestmt.close();
            }
            if(rs!=null)
            {
                rs.close();
            }
        }
    }
```

这里使用占位符?来初始化 SQL，然后通过不带参数的 addBatch 加入批处理中，最后还是通过 executeBatch 执行批处理操作。

执行结果：

```
影响的行数分别为:
1  1
```

上面演示了 Statement、PreparedStatement 批处理的使用，这里还要说明一下，批量执行 SQL 需要数据库的支持，有些数据库可能不支持。批量处理将多条 SQL 语句提交给数据库一块执行，效率高一些，但如果数据比较多，比如几万条 SQL，就需要分批次执行，例如 200 条执行一次，如果为了增加一致性，可以在批量处理的基础上增加事务。

5.1.7 事务处理

关系型数据库一般都支持事务。事务有四大特性：原子性、一致性、隔离性、持久性。

- 原子性：原子性是指事务包含的所有操作要么全部成功，要么全部失败回滚。例如转账，A 账户转给 B 账户，包含两个操作，将 A 账户的钱减去，然后将 B 账户加上对应的钱数，不可能 A 账户减了 B 账户没加上，也不可能 A 账户没减就给 B 账户加上了。两个操作要么都成功，要么都失败。
- 一致性：一致性是指事务必须使数据库从一个一致性状态变换到另一个一致性状态，也就是说一个事务执行之前和执行之后都必须处于一致性状态。假设账户 A 和账户 B 两者的钱加起来一共是 5000，那么不管 A 和 B 之间如何转账、转几次账，事务结束后两个用户的钱相加起来应该还得是 5000，这就是事务的一致性。这个就涉及隔离级别的问题了。
- 隔离性：隔离性是当多个用户并发访问数据库时，比如操作同一张表时，数据库为每一个用户开启的事务，不能被其他事务的操作所干扰，多个并发事务之间要相互隔离。
- 持久性：持久性是指一个事务一旦被提交了，那么对数据库中的数据的改变就是永久性的，即便是在数据库系统遇到故障的情况下也不会丢失提交事务的操作。

本例中使用 t_user 表的两个用户来模拟转账操作。目前李四账户有 888.88 元、张三账户有 666.66 元，让用户李四给用户张三转账 111.11 元，使两个账户都有 777.77 元。

id	name	age	money
1	张三	24	666.66
2	李四	25	888.88
3	王一	26	999.99
4	小明	27	555.55
5	小赵	28	333.33
6	小李	25	222.22

事务有两个结果：一是成功，二是回滚。在事务中，任何一个操作发生异常都会回滚。

```
public static void main( String[] args ) throws SQLException
{

    Connection conn=null;

    ResultSet rs=null;
    PreparedStatement prestmt=null;
    try{
        //注册驱动
        DriverManager.registerDriver(new com.mysql.jdbc.Driver());
        //通过注册的驱动获得连接对象 Connection
        conn=DriverManager.getConnection("jdbc:mysql://127.0.0.1:3306/
daodemodb?useUnicode=true&characterEncoding=UTF-8"
                + "&serverTimezone=UTC&useSSL=false","root","123456");
        //手动开启事务
        conn.setAutoCommit(false);
        String sql="update  t_user set money=money-? where id=?";
        prestmt=conn.prepareStatement(sql);

            prestmt.setDouble(1, 111.11);
            prestmt.setInt(2, 2);
            prestmt.addBatch();

            prestmt.setDouble(1, -111.11);
            prestmt.setInt(2, 1);
            prestmt.addBatch();

            //批处理
            prestmt.executeBatch();

            //提交事务
            conn.commit();

    }
    catch(SQLException e)
    {
        //事务回滚
        conn.rollback();
```

```
                System.out.println(e.getMessage());
                e.printStackTrace();
        }
        finally
        {
                //释放资源
                if(conn!=null)
                {
                    conn.close();
                }
                if(prestmt!=null)
                {
                    prestmt.close();
                }
                if(rs!=null)
                {
                    rs.close();
                }
        }
    }
```

上面代码只是做测试并未做金额是否满足转账要求检查，先使用“conn.setAutoCommit(false);”将自动提交设置为手动提交，默认是自动，然后批量执行两个 SQL 语句，在“conn.commit();”提交事务之前如果没有出现错误，执行结果会保存到数据库，一旦出现异常，就会执行“conn.rollback();”回滚操作。执行上面代码转账成功的输出结果如下：

id	name	age	money
1	张三	24	777.77
2	李四	25	777.77
3	王一	26	999.99
4	小明	27	555.55
5	小赵	28	333.33
6	小李	25	222.22

上面是提交成功的例子。为了演示事务回滚，可以在提交事务“conn.rollback();”之前制造一个异常“**int** a=1/0;”，然后执行，发现数据库的值并不会改变，并抛出了异常。

id	name	age	money
1	张三	24	777.77
2	李四	25	777.77
3	王一	26	999.99
4	小明	27	555.55
5	小赵	28	333.33
6	小李	25	222.22

5.2 Spring DAO 模块

Spring 的 DAO 模块提供了对 JDBC、Hibernate、MyBatis 等 DAO 层支持，本节介绍 DAO 模块对 JDBC 的支持。DAO 模块依赖 commons-dbcp.jar、commons-pool.jar。

5.2.1 JdbcDaoSupport 的使用

传统的 JDBC 需要创建连接、打开、执行 SQL、关闭连接这一系列步骤。Spring 框架对 JDBC 进行了封装，我们只需使用封装好的 JdbcTemplate 执行 SQL 语句。既然是 JdbcDaoSupport 的使用，为什么是使用 JdbcTemplate 呢？因为 JdbcDaoSupport 提供了 JdbcTemplate 对象，通过 JdbcTemplate 对象进行数据库操作。可以转到定义，查看 JdbcDaoSupport、JdbcTemplate 两个类的具体实现。我们通过下面的例子来了解 JdbcDaoSupport 的使用，这里还是使用 JDBC 章节的数据库 daodemodb 和表 t_user 信息。

第一步，根据 t_user 表信息准备 Model 类 User，定义 id、name、age、money 属性，并声明两个构造函数。

```
package com.demo.model;

public class User {

    @Override
    public String toString() {

        return "Id:"+this.getId()+" Name:"+this.getName()+" Age:"+ this.
getAge()+" Money:"+this.getMoney();
    }

    private int Id;
    private String Name;
    private int Age;
    private double Money;

    public User()
    {

    }

    public User(String name, int age, double money) {

        Name = name;
        Age = age;
        Money = money;
```

```
    }

    public int getId() {
        return Id;
    }

    public void setId(int id) {
        Id = id;
    }

    public String getName() {
        return Name;
    }

    public void setName(String name) {
        Name = name;
    }

    public int getAge() {
        return Age;
    }

    public void setAge(int age) {
        Age = age;
    }

    public double getMoney() {
        return Money;
    }

    public void setMoney(double money) {
        Money = money;
    }

}
```

第二步，定义接口类 IUserDAO，在接口中声明两个方法：QueryAllUser 方法属于查询操作，查询所有 User；AddUser 属于更新操作，新增 User。

```
package com.demo.model;

import java.util.*;

public interface IUserDAO {

    public List<User>QueryAllUser();
```

```
    public Boolean AddUser(User user);

}
```

第三步，就是 JdbcDaoSupport 的使用了。在下面的 SpringDAODemo 类中首先继承 JdbcDaoSupport，同时实现 IUserDAO 接口中的方法。JdbcDaoSupport 提供了 JdbcTemplate 对象，SpringDAODemo 继承了 JdbcDaoSupport，所以也就可以直接获取到 JdbcTemplate 对象，然后执行该对象的方法进行数据库操作。

```
package com.demo.model;

import java.util.*;

import org.springframework.beans.factory.BeanFactory;
import org.springframework.context.ApplicationContext;
import org.springframework.context.support.
ClassPathXmlApplicationContext;
import org.springframework.jdbc.core.support.JdbcDaoSupport;

public class SpringDAODemo extends JdbcDaoSupport implements IUserDAO {

    public static void main(String[] args) {
        ApplicationContext context=new ClassPathXmlApplicationContext(new
String[]{"ApplicationContext.xml"});
        BeanFactory factory=context;
        IUserDAO userDao=(IUserDAO)factory.getBean("userDao");
        User user=new User("JdbcDaoSupportTest",26,333.33);
        userDao.AddUser(user);
        List<User> list=userDao.QueryAllUser();
        for(User u:list)
        {
            System.out.println(u.toString());
        }
    }

    public List<User> QueryAllUser() {
        String sql="select id,name,age,money from t_user order by id desc";
        List<Map<String,Object>> list=getJdbcTemplate().queryForList(sql);
        List<User> userList=new ArrayList<User>();
        for(Map<String,Object> row:list)
        {
            User user=new User();
            user.setId((Integer)row.get("id"));
            user.setName((String)row.get("name"));
            user.setAge((Integer)row.get("age"));
            user.setMoney(Double.parseDouble(row.get("money").toString()));
            userList.add(user);
```

```
        }
        return userList;
    }

    public Boolean AddUser(User user) {
        String sql="insert into t_user (name,age,money) values (?,?,?)";
        int row=getJdbcTemplate().update(sql, new Object[]{user.getName(),
user.getAge(),user.getMoney()});
        if(row>0)
        {
            System.out.println("数据新增成功!");
            return true;
        }
        return false;
    }

}
```

第四步，配置属性。在上面的 main 方法中先通过上下文获取到 bean 对象，然后执行新增操作和查询操作。但是上面的代码并未看到数据库信息，这里还需要在 ApplicationContext.xml 中配置数据库信息，并为 JdbcDaoSupportDemo 设置数据源。为什么会有 dataSource 属性呢？因为 JdbcDaoSupport 中包含这个属性。

```
<?xml version="1.0" encoding="UTF-8"?>
<beans xmlns="http://www.springframework.org/schema/beans"
xmlns:context="http://www.springframework.org/schema/context"
xmlns:mvc="http://www.springframework.org/schema/mvc"
xmlns:xsi="http://www.w3.org/2001/XMLSchema-instance"
xmlns:aop="http://www.springframework.org/schema/aop"
xsi:schemaLocation="
http://www.springframework.org/schema/beans
http://www.springframework.org/schema/beans/spring-beans.xsd
http://www.springframework.org/schema/context
http://www.springframework.org/schema/context/spring-context.xsd
http://www.springframework.org/schema/mvc
http://www.springframework.org/schema/mvc/spring-mvc.xsd
http://www.springframework.org/schema/aop
http://www.springframework.org/schema/aop/spring-aop.xsd
">
<bean id="dataSource" class="org.apache.commons.dbcp.BasicDataSource">
  <property name="driverClassName">
        <value>com.mysql.jdbc.Driver</value>
  </property>
  <property name="url">
        <value>jdbc:mysql://127.0.0.1:3306/daodemodb</value>
  </property>
  <property name="username">
```

```
            <value>root</value>
        </property>
        <property name="password">
            <value>123456</value>
        </property>
    </bean>
    <bean id="userDao" class="com.demo.model.SpringDAODemo"
depends-on="dataSource">
        <property name="dataSource" ref="dataSource"></property>
    </bean>
    </beans>
```

第五步，pom.xml 中配置的依赖信息。例子是 Spring 中 DAO 的实现，而且是对 jdbc 的封装，所以包含 Spring 相关依赖和 jdbc 的相关依赖。

```
    <project xmlns="http://maven.apache.org/POM/4.0.0"
xmlns:xsi="http://www.w3.org/2001/XMLSchema-instance"
xsi:schemaLocation="http://maven.apache.org/POM/4.0.0
http://maven.apache.org/xsd/maven-4.0.0.xsd">
      <modelVersion>4.0.0</modelVersion>
      <groupId>com.demo</groupId>
      <artifactId>SpringDAO</artifactId>
      <version>0.0.1-SNAPSHOT</version>
      <properties>
        <project.build.sourceEncoding>UTF-8</project.build.sourceEncoding>
          <spring.version>5.0.0.RELEASE</spring.version>
      </properties>
      <dependencies>
           <dependency>
               <groupId>org.springframework</groupId>
               <artifactId>spring-core</artifactId>
               <version>${spring.version}</version>
           </dependency>
           <dependency>
               <groupId>org.springframework</groupId>
               <artifactId>spring-context</artifactId>
               <version>${spring.version}</version>
           </dependency>
           <dependency>
               <groupId>org.springframework</groupId>
               <artifactId>spring-jdbc</artifactId>
               <version>${spring.version}</version>
           </dependency>
           <dependency>
               <groupId>commons-dbcp</groupId>
               <artifactId>commons-dbcp</artifactId>
               <version>1.4</version>
```

```
        </dependency>
        <dependency>
            <groupId>commons-pool</groupId>
            <artifactId>commons-pool</artifactId>
            <version>1.6</version>
        </dependency>
        <!-- https://mvnrepository.com/artifact/mysql/mysql-connector-java
-->
        <dependency>
            <groupId>mysql</groupId>
            <artifactId>mysql-connector-java</artifactId>
            <version>5.1.6</version>
        </dependency>
      </dependencies>
</project>
```

运行 SpringDAODemo，可以发现数据库和打印结果中增加了一条记录。

```
信息: Loading XML bean definitions from class path r
数据新增成功!
Id:1 Name:JdbcDaoSupportTest Age:26 Money:333.33
```

5.2.2 MappingSqlQuery 的使用

在 JdbcDaoSupport 获取所有 User 的方法 QueryAllUser()中，使用 getJdbcTemplate().queryForList()返回的是 List<Map<String,Object>>类型，需要遍历转换成 Java 对象。问题来了，查询的不止这一个方法，可能以后会有条件查询的方法，每次都要把从数据库返回的 List<Map<String,Object>>类型的 List 转一遍，当然也可以专门写一个转换的方法，这样每次传 List<Map<String,Object>>类型的参数，然后返回 List<User>类型的值。其实还有一种方式，就是使用 MappingSqlQuery。MappingSqlQuery 是一个抽象类，需要实现它的方法 mapRow()。

第一步，实现 MappingSqlQuery 抽象类，这里在 UserMappingSqlQuery 类中实现了 mapRow()方法。

```
package com.demo.model;

import java.sql.ResultSet;
import java.sql.SQLException;

import org.springframework.jdbc.object.MappingSqlQuery;

public class UserMappingSqlQuery extends MappingSqlQuery<User>{

    @Override
    protected User mapRow(ResultSet rs, int rowNum) throws SQLException {
        User user=new User();
        user.setId((Integer)rs.getInt("id"));
        user.setName((String)rs.getString("name"));
        user.setAge((Integer)rs.getInt("age"));
```

```
            user.setMoney((Double)rs.getDouble("money"));
            return user;
        }

    }
```

第二步，UserMappingSqlQuery 类的使用。这里重写 SpringDAODemo 类中的 QueryAllUser()方法。UserMappingSqlQuery 需要传入 DataSource 和 sql，并执行 compile()编译。这里通过 getDataSource()获取的是 JdbcDaoSupport 的 DataSource 属性。如果有参数，可以使用 setParameters()设置参数，下面的代码为了演示设置参数，增加了 where 1=1 的查询条件。

```
    public List<User> QueryAllUser() {
        String sql="select id,name,age,money from t_user where ?";
        UserMappingSqlQuery userQuery=new UserMappingSqlQuery();
        userQuery.setDataSource(getDataSource());
        userQuery.setSql(sql);
        userQuery.setParameters(new SqlParameter(java.sql.Types.VARCHAR));
        userQuery.compile();
        return userQuery.execute(new Object[]{new String("1=1")});
    }
```

5.2.3　SqlUpdate 的使用

SqlUpdate 主要用来更新，可以设置参数。SqlUpdate 可以将某个功能模块化。通过下面的例子来了解一下 SqlUpdate 的使用。

```
    package com.demo.model;

    import javax.sql.DataSource;

    import org.springframework.beans.factory.BeanFactory;
    import org.springframework.context.ApplicationContext;
    import org.springframework.context.support.
ClassPathXmlApplicationContext;
    import org.springframework.jdbc.core.SqlParameter;
    import org.springframework.jdbc.object.SqlUpdate;

    public class UserSqlUpdate extends SqlUpdate{

       public static void main(String[] args) {

           ApplicationContext context=new ClassPathXmlApplicationContext(new
String[]{"ApplicationContext.xml"});
           BeanFactory factory=context;
           UserSqlUpdate userSqlUpdate=(UserSqlUpdate)factory.getBean
("userSqlUpdate");
           userSqlUpdate.updateUserMoney("小李",666.66);
       }

       public UserSqlUpdate(DataSource ds) {
           setDataSource(ds);
```

```
        setSql("update t_user set money=? where name=?");
        declareParameter(new SqlParameter(java.sql.Types.DOUBLE) );
        declareParameter(new SqlParameter(java.sql.Types.VARCHAR) );
        compile();
    }
    public Boolean updateUserMoney(String name,double money)
    {
        int row= update(new Object[]{new Double(money),new String(name)});
        if(row>0)
        {
            System.out.println("数据新增成功!");
            return true;
        }
        return false;
    }
}
```

在 ApplicationContext.xml 中配置 userSqlUpdate 对应的 bean 节点。

```
<bean id="userSqlUpdate" class="com.demo.model.UserSqlUpdate">
  <constructor-arg ref="dataSource" index="0"></constructor-arg>
</bean>
```

在上面的 UserSqlUpdate 中继承了 SqlUpdate，通过 setDataSource、setSql 分别设置 SqlUpdate 的数据源和要执行的 SQL。在 main 方法中将 name 为小李的 money 修改为 666.66，执行 main 方法之后会打印出“数据新增成功！”，数据库中小李的 money 改成了 666.66。

5.2.4 SqlFunction 的使用

SqlFunction 返回单一行的查询结果，默认返回 int，也可以重载返回其他类型。下面直接在 main 函数中使用。

```
public static void main(String[] args) {
    ApplicationContext context=new ClassPathXmlApplicationContext(new
String[]{"ApplicationContext.xml"});
    BeanFactory factory=context;
    BasicDataSource dataSource=(BasicDataSource)factory.
getBean("dataSource");
    SqlFunction sf=new SqlFunction(dataSource,"select count(1) from
t_user;");
    sf.compile();
    int count=sf.run();
    System.out.println("User Count:"+count);
}
```

5.3 Spring 事务管理

至于什么是事务，在 JDBC 章节已经介绍，这里就不再说了。JDBC 中有事务管理，Spring 使用 DataSourceTransactionManager 作为 JDBC 的事务管理者，同时把被管理的对象使用 TransactionProxyFactoryBean 配置。从名字也能猜出这里使用的设计模式是代理设计模式。这是一个事务代理 Bean，能够使用 IOC、AOP 等注入事务管理代码。在 JDBC 中介绍事务时用的是转账操作，这里为了更好理解，还是使用转账操作。

（1）在 IUserDAO 接口中增加转账方法 transfer。

```
public Boolean transfer(int fromUserId, int toUserId, float transferMoney);
```

（2）在 SpringDAODemo 类中实现 transfer。从 fromUserId 这个用户转账到 toUserId 这个用户，outInMoney 方法就是执行 SQL 更新数据库用户的 money。这里如果人为制造一个异常，把“int i=1/0; ”这行注释取消，就会在后面执行转账时自动回滚，转账不会成功。

```
public Boolean transfer(int fromUserId, int toUserId, float transferMoney){

        Boolean out= outInMoney(fromUserId,-transferMoney);
        //int i=1/0; //事务回滚
        Boolean in=outInMoney(toUserId,transferMoney);

        return out&in;
}
private Boolean outInMoney(int toUserId,float money)
{
        String sql="update  t_user set money=money+? where id=? ";
        int row=getJdbcTemplate().update(sql, new Object[]{money,toUserId});
        if(row>0)
        {
           return true;
        }
        return false;
}
```

（3）配置事务，在 ApplicationContext.xml 中配置事务管理对象 DataSourceTransaction-Manager，设置事务代理对象 TransactionProxyFactoryBean。

```
<bean id="transactionManager" class="org.springframework.jdbc.datasource.
DataSourceTransactionManager">
      <property name="dataSource" ref="dataSource"/>
</bean>
<!-- 配置业务层代理 -->
```

```
<bean id="userDaoProxy" class="org.springframework.transaction.
interceptor.TransactionProxyFactoryBean">
    <!-- 配置目标对象 -->
    <property name="target" ref="userDao"/>
    <!-- 注入事务管理器 -->
    <property name="transactionManager" ref="transactionManager"/>
    <!-- 注入事务的属性 -->
    <property name="transactionAttributes">
        <props>
            <prop key="transfer">PROPAGATION_REQUIRED</prop>
        </props>
    </property>
</bean>
```

（4）执行转账操作，在 main 方法中执行 transfer 方法，为用户 1、2 进行转账。如果在 transfer 方法中人为制造的异常被注释，是可以正常转账的，取消注释则转账失败，打印“事务回滚”。

```
public static void main(String[] args) {
     ApplicationContext context=new ClassPathXmlApplicationContext(new
String[]{"ApplicationContext.xml"});
      BeanFactory factory=context;
      IUserDAO userDao=(IUserDAO)factory.getBean("userDaoProxy");
      try
      {
         userDao.transfer(1, 2, 100);
      }
      catch(Exception e)
      {
          System.out.println("事务回滚");
      }
}
```

5.4 小结

本章主要介绍了 JDBC 的使用以及 Spring DAO 模块对 JDBC 的封装，Spring DAO 模块通过使用 JdbcDaoSupport、SqlUpdate 等对象隐藏了 Connection、Statement、ResultSet 等 JDBC API，而且可以比较灵活地配置事务管理。通过本章的学习，对后面学习 ORM 知识也打下基础。

第 6 章

MyBatis快速入门

在上一章节中，我们了解了 Spring 框架中的 DAO 模块，这一章我们将开始学习 ORM 框架中常用的一个框架——MyBatis。

本章主要涉及的知识点：

- MVC 框架：properties、environments、settings 等属性的配置。
- XML 映射文件：增删改查映射、参数、数据集的映射。
- 缓存：一级、二级缓存及自定义缓存。
- 动态 SQL：动态 SQL 的使用。
- 逆向工程：使用逆向工程自动生成配置文件
- 分页插件的使用：了解分页插件 pagehelper 的简单使用。

6.1 ORM 框架介绍

MyBatis 是一种 ORM 框架，在学习 MyBatis 之前我们需要了解 ORM 框架是什么以及为什么会有 ORM 框架。

6.1.1 ORM 框架简介

ORM（Object Relation Mapping，对象关系映射）是一种为了解决面向对象与关系数据库存在的互不匹配的现象的技术。通过使用描述对象和数据库之间映射的元数据，将程序中的对象自动持久化到关系数据库中。

在第 5 章学习的 JDBC 和 Spring 的 DAO 都可以进行数据持久化操作。那为什么还需要 ORM 框架呢？对于 SQL Server、MySQL、Oracle 等这些传统的数据库，基本都是关系型数据库，都是体现实体与实体之间的联系，在以前开发时，可能先根据需求设计数据库，然后写 Model 和业务逻辑，对于 Model 类基本都是和表的字段对应，而表中存的每条记录又和类的实例对象对应着。有了这个对照关系，就是能不能只在一边设计，在数据库设计表或在 VS 中设计 Model，然后直接生成另一边，这样就省了好多时间成本，于是有了 ORM。可以根据 Model 生成数据库，也可以由数据库生成 Model，Model、数据库是分离的，我们也可以根据 Model 生成不同类型（MySQL、SQL Server、Oracle）数据库，不同类型的数据库（MySQL、SQL Server、Oracle）也可以生成同样的 Model，而它们共同的纽带就是 Mapping。

6.1.2 MyBatis 框架介绍

Java 最常用的 ORM 中间件有 Hibernate 和 MyBatis。两个框架都有各自的优劣势，Hibernate 学习门槛不低，要精通门槛更高，为了能方便快速入门，这里主要介绍 MyBatis。有了 MyBatis 基础，对 ORM 思想有了进一步的认识之后再学习 Hibernate 也会相对容易一些。

我们经常也会看到 iBatis，其实它和 MyBatis 是同一个东西，MyBatis 本是 Apache 的一个开源项目 iBatis，2010 年这个项目由 Apache Software Foundation 迁移到了 Google Code，并且改名为 MyBatis。iBatis 一词来源于“internet”和“abatis”的组合，是一个基于 Java 的持久层框架。iBatis 提供的持久层框架包括 SQL Maps 和 Data Access Objects（DAO）。MyBatis 是 iBatis 的升级版，MyBatis 实现了接口绑定，使用更加方便，对象关系映射进行了改进，效率更高，采用功能强大的基于 OGNL 的表达式来消除其他元素。

MyBatis 是一款优秀的持久层框架，支持定制化 SQL、存储过程以及高级映射，MyBatis 避免了几乎所有的 JDBC 代码和手动设置参数以及获取结果集，可以使用简单的 xml 或注解来配置和映射原生信息，将接口和 Java 的 POJOs 映射成数据库的记录。

MyBatis 有以下特点：

（1）简单易学。没有任何第三方依赖，只需要两个 jar 包+几个 SQL 映射文件，通过文档和源代码，即可比较完全地掌握它的设计思路和实现。

（2）灵活。不会对应用程序或者数据库的现有设计强加任何影响。SQL 写在 xml 里面，便于统一管理和优化。通过 SQL 基本上可以实现不使用数据访问框架就能实现的所有功能。

（3）解除 SQL 与程序代码的耦合。通过提供 DAL 层，将业务逻辑和数据访问逻辑分离，使系统的设计更清晰、更易维护、更易于单元测试。

（4）提供映射标签，支持对象与数据库的 ORM 字段关系映射。

（5）提供对象关系映射标签，支持对象关系组建维护。

（6）提供 xml 标签，支持编写动态 SQL。

6.1.3 MyBatis 入门

使用 JDBC 时也有一些固定的步骤流程，而且每个步骤流程都或多或少对应着对象。有时候面试官也会经常问类似某某知识点（JDBC）有几大对象，如果掌握了它使用的步骤流程，就很容易记住有哪几个对象。MyBatis 的使用与 JDBC 类似，也有一些固定的步骤流程。所以学习 MyBatis 需要先了解一下它的使用流程。

1. 安装

使用 MyBatis 需要先引入 MyBatis 对应的依赖，同时需要操作数据库，所以也需要引入数据库对应的依赖，这里使用的是 MySQL 数据库连接依赖。

```
<!-- https://mvnrepository.com/artifact/org.mybatis/mybatis -->
<dependency>
    <groupId>org.mybatis</groupId>
    <artifactId>mybatis</artifactId>
```

```
        <version>3.4.6</version>
    </dependency>
    <!-- https://mvnrepository.com/artifact/mysql/mysql-connector-java -->
    <dependency>
        <groupId>mysql</groupId>
        <artifactId>mysql-connector-java</artifactId>
        <version>5.1.6</version>
    </dependency>
```

2. 构建 SqlSessionFactory

获取 SqlSessionFactory 与在第 3 章获取 Beanfactory 对象有点类似。Beanfactory 对象通过 ApplicationContext.xml 配置文件获取。每个基于 MyBatis 的应用都是以一个 SqlSessionFactory 实例为中心的。SqlSessionFactory 实例可以通过 SqlSessionFactoryBuilder 获得。SqlSessionFactoryBuilder 可以从 XML 配置文件或一个预先定制的 Configuration 实例构建出 SqlSessionFactory 实例。

从 XML 文件中构建 SqlSessionFactory 的实例非常简单，建议使用类路径下的资源文件进行配置，但是也可以使用任意的输入流（InputStream）实例，包括字符串形式的文件路径或者 file://的 URL 形式的文件路径。MyBatis 包含一个名叫 Resources 的工具类，包含一些实用方法，可使从 classpath 或其他位置加载资源文件更加容易。

```
    String resource = "mybatis-config.xml";
    //使用 MyBatis 提供的 Resources 类加载 mybatis 的配置文件（也加载关联的映射文件）
    Reader reader = Resources.getResourceAsReader(resource);
    //构建 sqlSession 的工厂
    SqlSessionFactory sessionFactory = new SqlSessionFactoryBuilder().
build(reader);
```

XML 配置文件（configuration XML）中包含了对 MyBatis 系统的核心设置，包含获取数据库连接实例的数据源（DataSource）和决定事务作用域和控制方式的事务管理器（TransactionManager），XML 配置文件会在后面的章节详细介绍。

3. 获取 SqlSession

既然有了 SqlSessionFactory，顾名思义，我们就可以从中获得 SqlSession 的实例了。SqlSession 完全包含了面向数据库执行 SQL 命令所需的所有方法。你可以通过 SqlSession 实例来直接执行已映射的 SQL 语句。这里先通过下面的例子来了解 MyBatis 的使用。

4. 实例演示

实例还是使用上一章节用到的 daodemodb 数据库、t_user 表以及 User 对象。这里就不再列出了。

（1）pom.xml 引入依赖库，主要引入 mybatis 依赖和 mysql-connector-javaMySql 数据库连接依赖。

```
    <!-- https://mvnrepository.com/artifact/org.mybatis/mybatis -->
```

```
    <dependency>
        <groupId>org.mybatis</groupId>
        <artifactId>mybatis</artifactId>
        <version>3.4.6</version>
    </dependency>
    <!-- https://mvnrepository.com/artifact/mysql/mysql-connector-java -->
    <dependency>
        <groupId>mysql</groupId>
        <artifactId>mysql-connector-java</artifactId>
        <version>5.1.6</version>
    </dependency>
```

（2）准备配置文件。SqlSessionFactory 对象是通过 XML 配置文件构建的，所以需要先准备配置文件，主要包含两个，一个是映射配置文件，另一个是 MyBatis 配置文件。配置信息具体含义可以先不用在意，后面会详细介绍。

UserMapper.xml 文件配置的是 SQL 语句与实体类的映射。

```
<?xml version="1.0" encoding="UTF-8" ?>
<!DOCTYPE mapper PUBLIC "-//mybatis.org//DTD Mapper 3.0//EN"
"http://mybatis.org/dtd/mybatis-3-mapper.dtd">
<mapper namespace="com.demo.mybatis.DBMapping.UserMapper">
    <select id="getUserList" resultType="com.demo.model.User">
        select * from t_user
    </select>
</mapper>
```

mybatis-config.xml 这里配置了数据库的连接信息以及 SQL 的 mappers。

```
<?xml version="1.0" encoding="UTF-8" ?>
<!DOCTYPE configuration
  PUBLIC "-//mybatis.org//DTD Config 3.0//EN"
  "http://mybatis.org/dtd/mybatis-3-config.dtd">
<configuration>
    <environments default="development">
        <environment id="development">
            <transactionManager type="JDBC" />
            <!-- 配置数据库连接信息 -->
            <dataSource type="POOLED">
                <property name="driver" value="com.mysql.jdbc.Driver" />
                <property name="url" value="jdbc:mysql://127.0.0.1:3306/
daodemodb"/>
                <property name="username" value="root" />
                <property name="password" value="123456" />
            </dataSource>
        </environment>
    </environments>
    <!-- 程序中所用到sql映射文件都在这里列出，这些映射sql都被MyBatis管理 -->
    <mappers>
```

```
        <mapper resource="com/demo/mybatis/DBMapping/UserMapper.xml" />
    </mappers>
</configuration>
```

（3）通过准备好的配置文件构建 SqlSessionFactory，获取到能执行映射文件中 SQL 的 SqlSession，然后执行 SQL 输出 t_user 表中所有的用户信息。

```
public class BasicDemo {
    public static void main( String[] args ) throws IOException
    {
         String resource = "mybatis-config.xml";
         //使用MyBatis提供的Resources类加载mybatis的配置文件（也加载关联的映射文件）
         Reader reader = Resources.getResourceAsReader(resource);
         //构建sqlSession的工厂
         SqlSessionFactory sessionFactory = new SqlSessionFactoryBuilder().
build(reader);
         //创建能执行映射文件中sql的sqlSession
         SqlSession session = sessionFactory.openSession();
         String statement="com.demo.mybatis.DBMapping.UserMapper.
getUserList";
         List<User>users=session.selectList(statement);
         for (User u: users) {
           System.out.println(u.toString());
         }
    }
}
```

通过上面的一个查询用户表信息的 demo，基本了解了 MyBatis 的使用过程，下面详细了解一下 XML 配置的详细内容。

6.2 XML 配置

MyBatis 的配置文件包含了会深深影响 MyBatis 行为的设置(settings)和属性(properties)信息。在上面 MyBatis 实例中 mybatis-config.xml 只是配置了 environments 信息，比较简单，本节详细了解一下 MyBatis 的配置。

6.2.1 properties 属性

属性都是可外部配置且可动态替换的，既可以在典型的 Java 属性文件中配置，亦可通过 properties 元素的子元素来传递。上面<dataSource>配置中的属性就可以先配置在 properties 元素中。这里在 resource 包中增加了 database.properties 文件，在 database.properties 设置了 driver 和 url 属性。

```
driver=com.mysql.jdbc.Driver
url=jdbc:mysql://127.0.0.1:3306/daodemodb
```

在 mybatis-config.xml 中增加 properties 元素，这里通过 resource 引入 database.properties 文件中的属性，同时设置 username 和 password 两个属性。

```
<properties resource="database.properties">
  <property name="username" value="root"/>
  <property name="password" value="123456"/>
</properties>
```

有了这些属性之后，就可以将上面 dataSource 元素的配置进行修改，不用写成固定值，而是写成属性值，这样做有什么好处呢？其实有些属性可能不止一个地方会用到，一个地方设置，多个地方引用，这样后续维护也会更方便一些。

```
<?xml version="1.0" encoding="UTF-8" ?>
<!DOCTYPE configuration
  PUBLIC "-//mybatis.org//DTD Config 3.0//EN"
  "http://mybatis.org/dtd/mybatis-3-config.dtd">
<configuration>
    <properties resource="database.properties">
      <property name="username" value="root"/>
      <property name="password" value="123456"/>
    </properties>
   <environments default="development">
       <environment id="development">
           <transactionManager type="JDBC" />
           <!-- 配置数据库连接信息 -->
           <dataSource type="POOLED">
               <property name="driver" value="${driver}" />
               <property name="url" value="${url}" />
               <property name="username" value="${username}" />
               <property name="password" value="${password}" />
           </dataSource>
       </environment>
    </environments>
    <!-- 程序中所用到 sql 映射文件都在这里列出，这些映射 sql 都被 MyBatis 管理 -->
    <mappers>
       <mapper resource="com/demo/mybatis/DBMapping/UserMapper.xml" />
    </mappers>
</configuration>
```

这里 username 和 password 将会由 properties 元素中设置的相应值来替换。driver 和 url 属性将会由 database.properties 文件中对应的值来替换，这样也为配置提供了诸多灵活选择。

如果属性在不止一个地方进行了配置，那么 MyBatis 将按照下面的顺序来加载：

（1）在 properties 元素体内指定的属性首先被读取。

（2）根据 properties 元素中的 resource 属性读取类路径下属性文件或根据 url 属性指定的路径读取属性文件，并覆盖已读取的同名属性。

（3）读取作为方法参数传递的属性，并覆盖已读取的同名属性。

因此，通过方法参数传递的属性具有最高优先级，resource/url 属性中指定的配置文件次之，最低优先级的是 properties 属性中指定的属性。

6.2.2 settings

settings 是 MyBatis 中极为重要的调整设置，它们会改变 MyBatis 的运行时行为。表 6-1 给出设置中各项的意图、默认值等。

表 6-1 settings 各项的意图、默认值

设置参数	描　述	有 效 值	默 认 值
cacheEnabled	全局地开启或关闭配置文件中的所有映射器已经配置的任何缓存	true | false	true
lazyLoadingEnabled	延迟加载的全局开关。当开启时，所有关联对象都会延迟加载。特定关联关系中可通过设置 fetchType 属性来覆盖该项的开关状态	true | false	false
aggressiveLazyLoading	当开启时，任何方法的调用都会加载该对象的所有属性。否则，每个属性会按需加载（参考 lazyLoadTriggerMethods）	true | false	false（ true in ≤ 3.4.1）
multipleResultSetsEnabled	是否允许单一语句返回多结果集（需要兼容驱动）	true | false	true
useColumnLabel	使用列标签代替列名。不同的驱动在这方面会有不同的表现，具体可参考相关驱动文档，或通过测试这两种不同的模式来观察所用驱动的结果	true | false	true
useGeneratedKeys	允许 JDBC 支持自动生成主键，需要驱动兼容。如果设置为 true，那么这个设置强制使用自动生成主键，尽管一些驱动不能兼容但仍可正常工作（比如 Derby）	true | false	False
autoMappingBehavior	指定 MyBatis 应如何自动映射列到字段或属性。NONE 表示取消自动映射；PARTIAL 只会自动映射没有定义嵌套结果集映射的结果集；FULL 会自动映射任意复杂的结果集（无论是否嵌套）	NONE，PARTIAL，FULL	PARTIAL

（续表）

设置参数	描　述	有 效 值	默 认 值
autoMappingUnknown-ColumnBehavior	指定发现自动映射目标未知列（或者未知属性类型）的行为。 ● NONE：不做任何反应 ● WARNING：输出提醒日志（'org.apache.ibatis.session.AutoMapping-UnknownColumnBehavior'的日志等级必须设置为 WARN） ● FAILING：映射失败（抛出 SqlSession-Exception）	NONE，WARNING，FAILING	NONE
defaultExecutorType	配置默认的执行器。SIMPLE 就是普通的执行器；REUSE 执行器会重用预处理语句（prepared statements）；BATCH 执行器将重用语句并执行批量更新	SIMPLE REUSE BATCH	SIMPLE
defaultStatementTimeout	设置超时时间，决定驱动等待数据库响应的秒数	任意正整数	Not Set (null)
defaultFetchSize	为驱动的结果集获取数量（fetchSize）设置一个提示值。此参数只可以在查询设置中被覆盖	任意正整数	Not Set (null)
safeRowBoundsEnabled	允许在嵌套语句中使用分页（RowBounds）。如果允许使用就设置为 false	true | false	False
safeResultHandlerEnabled	允许在嵌套语句中使用分页（ResultHandler）。如果允许使用就设置为 false	true | false	True
mapUnderscoreToCamelCase	是否开启自动驼峰命名规则（camel case）映射，即从经典数据库列名 A_COLUMN 到经典 Java 属性名 aColumn 的类似映射	true | false	False
localCacheScope	MyBatis 利用本地缓存机制（Local Cache）防止循环引用（circular references）和加速重复嵌套查询。默认值为 SESSION，这种情况下会缓存一个会话中执行的所有查询。若设置值为 STATEMENT，本地会话仅用在语句执行上，对相同 SqlSession 的不同调用将不会共享数据	SESSION | STATEMENT	SESSION
jdbcTypeForNull	当没有为参数提供特定的 JDBC 类型时，为空值指定 JDBC 类型。某些驱动需要指定列的 JDBC 类型，多数情况下直接用一般类型即可，比如 NULL、VARCHAR 或 OTHER	JdbcType 常量	OTHER

（续表）

设置参数	描　述	有 效 值	默 认 值
lazyLoadTriggerMethods	指定哪个对象的方法触发一次延迟加载	用逗号分隔的方法列表	equals,clone，hashCode，toString
defaultScriptingLanguage	指定动态 SQL 生成的默认语言	一个类型别名或完全限定类名	org.apache.ibatis.scripting.xmltags.XMLLanguageDriver
defaultEnumTypeHandler	指定 Enum 使用的默认 TypeHandler（从 3.4.5 开始）	一个类型别名或完全限定类名	org.apache.ibatis.type.EnumTypeHandler
callSettersOnNulls	指定当结果集中值为 null 的时候是否调用映射对象的 setter（map 对象时为 put）方法，这对于有 Map.keySet()依赖或 null 值初始化的时候是有用的。注意基本类型（int、boolean 等）是不能设置成 null 的	true | false	false
returnInstanceForEmptyRow	当返回行的所有列都是空时，MyBatis 默认返回 null。当开启这个设置时，MyBatis 会返回一个空实例。注意，它也适用于嵌套的结果集（i.e. collectioin and association）（从 3.4.2 开始）	true | false	false
logPrefix	指定 MyBatis 增加到日志名称的前缀	任何字符串	Not set
proxyFactory	指定 MyBatis 创建具有延迟加载能力的对象所用到的代理工具	CGLIB | JAVASSIST	JAVASSIST（MyBatis 3.3 or above）
vfsImpl	指定 VFS 的实现	自定义 VFS 的实现的类全限定名，以逗号分隔	Not set
useActualParamName	允许使用方法签名中的名称作为语句参数名称。为了使用该特性，你的工程必须采用 Java 8 编译，并且加上-parameters 选项（从 3.4.1 开始）	true | false	true
configurationFactory	指定一个提供 Configuration 实例的类。这个被返回的 Configuration 实例用来加载被反序列化对象的懒加载属性值。这个类必须包含一个签名方法 static Configuration getConfiguration()（从 3.2.3 版本开始）	类型别名或者全类名	Not set

一个配置完整的 settings 元素的示例如下：

```
<settings>
  <setting name="cacheEnabled" value="true"/>
  <setting name="lazyLoadingEnabled" value="true"/>
  <setting name="multipleResultSetsEnabled" value="true"/>
  <setting name="useColumnLabel" value="true"/>
  <setting name="useGeneratedKeys" value="false"/>
  <setting name="autoMappingBehavior" value="PARTIAL"/>
  <setting name="autoMappingUnknownColumnBehavior" value="WARNING"/>
  <setting name="defaultExecutorType" value="SIMPLE"/>
  <setting name="defaultStatementTimeout" value="25"/>
  <setting name="defaultFetchSize" value="100"/>
  <setting name="safeRowBoundsEnabled" value="false"/>
  <setting name="mapUnderscoreToCamelCase" value="false"/>
  <setting name="localCacheScope" value="SESSION"/>
  <setting name="jdbcTypeForNull" value="OTHER"/>
  <setting name="lazyLoadTriggerMethods" value="equals,clone,hashCode,
toString"/>
</settings>
```

6.2.3 typeAliases

类型别名是为 Java 类型设置一个短的名字。它只和 XML 配置有关，存在的意义仅在于用来减少类完全限定名的冗余。在 6.1.3 节的 UserMapper.xml 文件中，设置 resultType 时用的是 User 类完全限定名。

```
<?xml version="1.0" encoding="UTF-8" ?>
<!DOCTYPE mapper PUBLIC "-//mybatis.org//DTD Mapper 3.0//EN"
"http://mybatis.org/dtd/mybatis-3-mapper.dtd">
<mapper namespace="com.demo.mybatis.DBMapping.UserMapper">
  <select id="getUserList" resultType="com.demo.model.User">
    select * from t_user
  </select>
</mapper>
```

这里在 mybatis-config.xml 中使用 typeAliases 设置 com.demo.model.User 的别名为 User。

```
<typeAliases>
  <typeAlias type="com.demo.model.User" alias="User"/>
</typeAliases>
```

有了别名之后，在 UserMapper.xml 中用到 com.demo.model.User 的都可以用别名代替。下面的 xml 配置中将 resultType=com.demo.model.User 改成了 resultType=User。有一点需要注意，别名不区分大小写，resultType=USER 或 user 都是可以的。

```
<?xml version="1.0" encoding="UTF-8" ?>
<!DOCTYPE mapper PUBLIC "-//mybatis.org//DTD Mapper 3.0//EN"
```

```
"http://mybatis.org/dtd/mybatis-3-mapper.dtd">
    <mapper namespace="com.demo.mybatis.DBMapping.UserMapper">
        <select id="getUserList" resultType="User">
            select * from t_user
        </select>
    </mapper>
```

通过设置别名，减少了类完全限定名的冗余，但还有一个问题，一个项目中 Java Bean 会很多，每个 Java Bean 都要配置一次，有些麻烦，有没有进一步简化的方法呢？答案肯定是有的。可以指定一个包名，MyBatis 会在包名下面搜索需要的 Java Bean。在没有注解的情况下，会使用 Bean 的非限定类名来作为它的别名。

```
<typeAliases>
        <package name="com.demo.model"/>
</typeAliases>
```

若有注解，则别名为其注解值。这里在 User 类设置别名为 users，在设置 resultType 时可以使用 users 别名。

```
@Alias("users")
public class User {

<?xml version="1.0" encoding="UTF-8" ?>
<!DOCTYPE mapper PUBLIC "-//mybatis.org//DTD Mapper 3.0//EN"
"http://mybatis.org/dtd/mybatis-3-mapper.dtd">
<mapper namespace="com.demo.mybatis.DBMapping.UserMapper">
    <select id="getUserList" resultType="users">
        select * from t_user
    </select>
</mapper>
```

6.2.4　typeHandlers

1. 默认类型处理器

无论是 MyBatis 在预处理语句（PreparedStatement）中设置一个参数，还是从结果集中取出一个值，都会用类型处理器将获取的值以合适的方式转换成 Java 类型。这个转换就需要使用 typeHandlers。表 6-2 就是一些它默认自带的类型处理器。

表 6-2　typeHandlers 默认自带的类型处理器

类型处理器	Java 类型	JDBC 类型
BooleanTypeHandler	java.lang.Boolean, boolean	数据库兼容的 BOOLEAN
ByteTypeHandler	java.lang.Byte, byte	数据库兼容的 NUMERIC 或 BYTE
ShortTypeHandler	java.lang.Short, short	数据库兼容的 NUMERIC 或 SHORT INTEGER
IntegerTypeHandler	java.lang.Integer, int	数据库兼容的 NUMERIC 或 INTEGER

（续表）

类型处理器	Java 类型	JDBC 类型
LongTypeHandler	java.lang.Long, long	数据库兼容的 NUMERIC 或 数据库兼容的 LONG INTEGER
FloatTypeHandler	java.lang.Float, float	数据库兼容的 NUMERIC 或 FLOAT
DoubleTypeHandler	java.lang.Double, double	数据库兼容的 NUMERIC 或 DOUBLE
BigDecimalTypeHandler	java.math.BigDecimal	数据库兼容的 NUMERIC 或 DECIMAL
StringTypeHandler	java.lang.String	CHAR, VARCHAR
ClobTypeHandler	java.lang.String	CLOB, LONGVARCHAR
NStringTypeHandler	java.lang.String	NVARCHAR, NCHAR
NClobTypeHandler	java.lang.String	NCLOB
ByteArrayTypeHandler	byte[]	数据库兼容的字节流类型
BlobTypeHandler	byte[]	BLOB, LONGVARBINARY
DateTypeHandler	java.util.Date	TIMESTAMP
DateOnlyTypeHandler	java.util.Date	DATE
TimeOnlyTypeHandler	java.util.Date	TIME
SqlTimestampTypeHandler	java.sql.Timestamp	TIMESTAMP
SqlDateTypeHandler	java.sql.Date	DATE
SqlTimeTypeHandler	java.sql.Time	TIME
EnumTypeHandler	Enumeration Type	VARCHAR，任何兼容的字符串类型，存储枚举的名称（而不是索引）
EnumOrdinalTypeHandler	Enumeration Type	任何兼容的 NUMERIC 或 DOUBLE 类型，存储枚举的索引（而不是名称）
InstantTypeHandler	java.time.Instant	TIMESTAMP
LocalDateTimeTypeHandler	java.time.LocalDateTime	TIMESTAMP
LocalDateTypeHandler	java.time.LocalDate	DATE
LocalTimeTypeHandler	java.time.LocalTime	TIME
OffsetDateTimeTypeHandler	java.time.OffsetDateTime	TIMESTAMP
OffsetTimeTypeHandler	java.time.OffsetTime	TIME
ZonedDateTimeTypeHandler	java.time.ZonedDateTime	TIMESTAMP
YearTypeHandler	java.time.Year	INTEGER
MonthTypeHandler	java.time.Month	INTEGER
YearMonthTypeHandler	java.time.YearMonth	VARCHAR or LONGVARCHAR
JapaneseDateTypeHandler	java.time.chrono.JapaneseDate	DATE

针对上面默认自带的类型处理器，我们以 EnumTypeHandler、EnumOrdinalTypeHandler 为例演示自带类型处理器的使用。二者的区别是 EnumTypeHandler 直接存储 name 值，而 EnumOrdinalTypeHandler 会存储 enum 类型里的序号值，此时数据库表字段一般用 int 类型的处

理。我们来看一下 EnumTypeHandler、EnumOrdinalTypeHandler 的使用。

这里为了演示枚举类型，首先增加一个枚举类 UserState。

```
package com.demo.model;
public enum UserState {
    DISABLED(0),
    AVAILABLE(1);
    private int status;
    UserState(int status)
    {
        this.status=status;
    }
    public int getStatus()
    {
        return status;
    }
}
```

在 User 类中增加一个 UserState 类型的属性 Status，与之对应，在数据库中也增加一个字段 status，由于 EnumTypeHandler 直接存储 name 值，而 EnumOrdinalTypeHandler 会存储 enum 类里的序号值字段类型，为了方便把字段设置成了 varchar 类型。

User 类增加属性：

```
private UserState Status;

public UserState getStatus() {
    return Status;
}

public void setStatus(UserState status) {
    Status = status;
}
```

t_user 表增加 status 列：

```
ALTER TABLE `daodemodb`.`t_user`
ADD COLUMN `states` VARCHAR(45) NULL AFTER `money`;
```

下面要面对的问题是怎么将 User 类中的 UserState 类型的属性 Status 与数据库 t_user 表的 status 对应上，两个互相转换的问题。在 UserMapper.xml 中增加了两个操作，一个是新增用户并返回用户自增 id，一个是根据用户 id 查询用户。

```
<?xml version="1.0" encoding="UTF-8" ?>
<!DOCTYPE mapper PUBLIC "-//mybatis.org//DTD Mapper 3.0//EN"
"http://mybatis.org/dtd/mybatis-3-mapper.dtd">
<mapper namespace="com.demo.mybatis.DBMapping.UserMapper">
    <select id="getUserList" resultMap="userResult">
        select * from t_user
```

```
        </select>
        <insert id="addUser" parameterType="user">
          <selectKey resultType="java.lang.Integer" order="AFTER"
keyProperty="id">
              SELECT LAST_INSERT_ID()
          </selectKey>
            insert into t_user(name,age,money,status) values
(#{name},#{age},#{money},#{status,typeHandler=org.apache.ibatis.type.EnumTyp
eHandler})
        </insert>
        <resultMap type="user" id="userResult">
            <result column="id" property="id"/>
            <result column="name" property="name"/>
            <result column="age" property="age"/>
            <result column="money" property="money"/>
            <result column="status" property="status"
typeHandler="org.apache.ibatis.type.EnumTypeHandler"/>
        </resultMap>
        <select id="getUser" parameterType="int"
            resultType="User" resultMap="userResult">
            select * from t_user where id=#{id}
        </select>
    </mapper>
```

这里主要了解枚举的使用，涉及 resultMap 的一些知识可以先略过，后面会对 select、resultMap 这些做进一步介绍。在上面 UserMapper.xml 中使用 typeHandler 来指定枚举类型，由于 typeHandler=org.apache.ibatis.type.EnumTypeHandler，因此在新增用户时数据库保存的是 UserState 的 name。如果想保存枚举的索引，只需把 typeHandler 改为 EnumOrdinalTypeHandler。在下面的 mian 方法中，首先新增一个 user 并获取到对应的自增 id，根据 id 再查询出 user，打印出来。有一点要特别注意，别忘记执行 session.commit()，否则虽然会查询到用户自增 id，但数据库中并不存在数据。

```
    public static void main( String[] args ) throws IOException
    {
        String resource = "mybatis-config.xml";
         //使用MyBatis提供的Resources类加载mybatis的配置文件(它也加载关联的映射文件)
         Reader reader = Resources.getResourceAsReader(resource);
         //构建 sqlSession 的工厂
         SqlSessionFactory sessionFactory = new SqlSessionFactoryBuilder().
build(reader);
         //创建能执行映射文件中 sql 的 sqlSession
         SqlSession session = sessionFactory.openSession();
         String statement="com.demo.mybatis.DBMapping.UserMapper.addUser";
         User user=new User("usermode",22,222.22,UserState.AVAILABLE);
         session.insert(statement,user);
         session.commit();
```

```
        System.out.println(user.getId());
        statement="com.demo.mybatis.DBMapping.UserMapper.getUser";
        user=session.selectOne(statement, user.getId());
        System.out.println(user.toString());
    }
```

输出结果如下：

```
46
Id:46 Name:usermode Age:22 Money:222.22 Status:AVAILABLE
```

2. 自定义处理器

有时候 MyBatis 默认的类型转换器并不能满足需求，我们也可以自定义。MyBatis 提供了 org.apache.ibatis.type.BaseTypeHandler 类，用于我们自己扩展类型转换器，上面的 EnumTypeHandler 和 EnumOrdinalTypeHandler 也都实现了这个接口。这里自定义了 CusEnumStatusHandler 类，实现了 BaseTypeHandler 接口。这里采用 EnumOrdinalTypeHandler 模式保存数字，在用的时候直接引用。

```
package com.demo.model;
import java.sql.CallableStatement;
import java.sql.PreparedStatement;
import java.sql.ResultSet;
import java.sql.SQLException;
import org.apache.ibatis.type.BaseTypeHandler;
import org.apache.ibatis.type.JdbcType;

public class CusEnumStatusHandler extends BaseTypeHandler<UserState>{

    @Override
    public void setNonNullParameter(PreparedStatement ps, int i, UserState
parameter, JdbcType jdbcType)
            throws SQLException {
        ps.setInt(i, parameter.getStatus());
    }

    @Override
    public UserState getNullableResult(ResultSet rs, String columnName)
throws SQLException {
        return UserState.fromValue(rs.getInt(columnName));
    }

    @Override
    public UserState getNullableResult(ResultSet rs, int columnIndex) throws
SQLException {

        return UserState.fromValue(rs.getInt(columnIndex));
    }
```

```
    @Override
    public UserState getNullableResult(CallableStatement cs, int columnIndex)
throws SQLException {

        return UserState.fromValue(cs.getInt(columnIndex));
    }

}
```

在上面的代码中主要调用了 UserState 中的 fromValue 方法，下面是 fromValue 方法的实现。

```
public static UserState fromValue(int value)
{
    for(UserState userState:UserState.values())
    {
        if(userState.status==value)
        {
            return userState;
        }
    }
    throw new IllegalArgumentException("Cannot create evalue from value: "
+ value + "!");
}
```

定义好之后只需将 UserMapper.xml 中 resultMap 节点中的 typeHandler 设置为自定义类型转换器类即可。此时再执行 main 方法中的新增查询，一样可以查出结果。

```
<resultMap type="user" id="userResult">
    <result column="id" property="id"/>
    <result column="name" property="name"/>
    <result column="age" property="age"/>
    <result column="money" property="money"/>
    <result column="status" property="status"
typeHandler="com.demo.model.CusEnumStatusHandler"/>
</resultMap>
```

6.2.5 配置环境（environments）

在项目开发过程中往往有几套环境，例如开发、测试和生产环境，而每个环境可能会有不同的配置，或者共享相同 Schema 的多个生产数据库。MyBatis 可以配置成适应多种环境，这种机制有助于将 SQL 映射应用于多种数据库之中。不过尽管可以配置多个环境，每个 SqlSessionFactory 实例只能选择其一。如果想连接两个数据库，就需要创建两个 SqlSessionFactory 实例，每个数据库对应一个。

为了指定创建哪种环境，只要将它作为可选的参数传递给 SqlSessionFactoryBuilder 即可，其中 environment、properties 两个参数是可选，如果忽略了环境参数，那么默认环境将会被加载。

```
SqlSessionFactory factory = new SqlSessionFactoryBuilder().build(reader,
environment, properties);
```

环境元素如何配置呢？它又包含哪些属性信息呢？下面是 6.1.3 节配置的环境。

```
<environments default="development">
    <environment id="development">
        <transactionManager type="JDBC" />
        <!-- 配置数据库连接信息 -->
        <dataSource type="POOLED">
            <property name="driver" value="${driver}" />
            <property name="url" value="${url}" />
            <property name="username" value="${username}" />
            <property name="password" value="${password}" />
        </dataSource>
    </environment>
</environments>
```

MyBatis 支持多环境配置，所以 environments 元素可以配置多个 environment 子元素，每一个子元素使用 id 属性区分，在 environments 中通过 default 属性指定默认环境配置。在每个子元素 environment 中又包含两个子元素：事务管理器（transactionManager）和数据源（dataSource）。

1. 事务管理器（transactionManager）

在 MyBatis 中有两种类型的事务管理器（也就是 type="[JDBC|MANAGED] "）：

- JDBC：这个配置就是直接使用了 JDBC 的提交和回滚设置，它依赖于从数据源得到的连接来管理事务作用域。上面的配置就是使用的 JDBC 类型。
- MANAGED：这个配置几乎没做什么。它从来不提交或回滚一个连接，而是让容器来管理事务的整个生命周期（比如 JEE 应用服务器的上下文）。默认情况下，它会关闭连接，然而一些容器并不希望这样，因此需要将 closeConnection 属性设置为 false 来阻止它默认的关闭行为。例如:

```
<transactionManager type="MANAGED">
    <property name="closeConnection" value="false"/>
</transactionManager>
```

如果使用 Spring + MyBatis，就没有必要配置事务管理器，因为 Spring 模块会使用自带的管理器来覆盖前面的配置。

2. 数据源（dataSource）

dataSource 元素使用标准的 JDBC 数据源接口来配置 JDBC 连接对象的资源。有三种内建

的数据源类型（也就是 type="[UNPOOLED|POOLED|JNDI]"）。上面例子设置的 type=POOLED。

- UNPOOLED：这个数据源的实现只是每次被请求时打开和关闭连接。虽然有点慢，但对于在数据库连接可用性方面没有太高要求的简单应用程序来说，是一个很好的选择。不同的数据库在性能方面的表现也是不一样的，对于某些数据库来说，使用连接池并不重要，这个配置就很适合这种情形。UNPOOLED 类型的数据源仅仅需要配置以下 5 种属性：
  - driver：这是 JDBC 驱动的 Java 类的完全限定名（并不是 JDBC 驱动中可能包含的数据源类）。
  - url：这是数据库的 JDBC URL 地址。
  - username：登录数据库的用户名。
  - password：登录数据库的密码。
  - defaultTransactionIsolationLevel：默认的连接事务隔离级别。
- POOLED：这种数据源的实现利用“池”的概念将 JDBC 连接对象组织起来，避免了创建新的连接实例时所必需的初始化和认证时间。这是一种使得并发 Web 应用快速响应请求的流行处理方式。
  除了上述提到 UNPOOLED 下的属性外，还有更多属性用来配置 POOLED 的数据源：
  - poolMaximumActiveConnections：在任意时间可以存在的活动（也就是正在使用）连接数量，默认值为 10。
  - poolMaximumIdleConnections：任意时间可能存在的空闲连接数。
  - poolMaximumCheckoutTime：在被强制返回之前，池中连接被检出（checked out）时间，默认值为 20000 毫秒（20 秒）。
  - poolTimeToWait：这是一个底层设置，如果获取连接花费了相当长的时间，连接池会打印状态日志并重新尝试获取一个连接（避免在误配置的情况下一直安静地失败），默认值为 20000 毫秒（20 秒）。
  - poolMaximumLocalBadConnectionTolerance：这是一个关于坏连接容忍度的底层设置，作用于每一个尝试从缓存池获取连接的线程。如果这个线程获取到的是一个坏的连接，数据源就会允许这个线程尝试重新获取一个新的连接，但是这个重新尝试的次数不应该超过 poolMaximumIdleConnections 与 poolMaximumLocalBadConnection-Tolerance 之和。默认值为 3（新增于 3.4.5）。
  - poolPingQuery：发送到数据库的侦测查询，用来检验连接是否正常工作并准备接受请求。默认是“NO PING QUERY SET”，这会导致多数数据库驱动失败时带有一个恰当的错误消息。
  - poolPingEnabled：是否启用侦测查询。若开启，需要设置 poolPingQuery 属性为一个可执行的 SQL 语句（最好是一个速度非常快的 SQL 语句），默认值为 false。
  - poolPingConnectionsNotUsedFor：配置 poolPingQuery 的频率。可以被设置为和数据库连接超时时间一样来避免不必要的侦测，默认值为 0（所有连接每一时刻都被侦测，当然仅当 poolPingEnabled 为 true 时适用）。

- JNDI：这个数据源的实现是为了能在 EJB 或应用服务器这类容器中使用，容器可以集中或在外部配置数据源，然后放置一个 JNDI 上下文的引用。这种数据源配置只需要两个属性：
  - initial_context：这个属性用来在 InitialContext 中寻找上下文（即 initialContext.lookup(initial_context)）。这是一个可选属性，如果忽略，那么 data_source 属性将会直接从 InitialContext 中寻找。
  - data_source：这是引用数据源实例位置的上下文的路径。提供了 initial_context 配置时会在其返回的上下文中进行查找，没有提供时则直接在 InitialContext 中查找。

6.2.6　映射器（mappers）

在 UserMapper.xml 文件配置中还有一个元素未介绍，那就是 mappers 元素，实体类与数据库表的纽带就是通过该元素。mappers 元素包含若干 mapper 子元素，子元素告诉 MyBatis 到哪里去找映射文件。这里可以使用相对于类路径的资源引用，或完全限定资源定位符（包括 file:///的 URL），或类名和包名等。在 UserMapper.xml 中使用的是相对于类路径的资源引用。

```
<mappers>
    <mapper resource="com/demo/mybatis/DBMapping/UserMapper.xml" />
</mappers>
```

mapper 元素也可以通过 url 属性指定一个 Mapper.xml 文件。

```
<mappers>
    <mapper url="file:///C:/Users/admin/Desktop/UserMapper.xml"/>
</mappers>
```

使用 mappers 映射器配置会告诉 MyBatis 去哪里找映射文件，剩下的细节就应该是每个 SQL 映射文件了。

6.3　XML 映射文件

映射是 ORM 中最主要的部分，MyBatis 的真正强大在于它的映射语句，也是它的魔力所在。由于它异常强大，映射器的 XML 文件就显得相对简单。如果拿它跟具有相同功能的 JDBC 代码进行对比，你会立即发现省掉了将近 95%的代码。MyBatis 就是针对 SQL 构建的，并且比普通的方法做的更好。SQL 操作大部分都是增、删、改、查，所以可以参照 SQL 操作来了解 MyBatis 对应的实现。SQL 映射文件有很少的几个顶级元素：cache、cache-ref、resultMap、sql、insert、update、delete、select。

6.3.1　查询元素 select

在上面 UserMapper.xml 中，有两个地方使用到了 select 元素，一个是查询 Tt_user 表中所有 user，另一个是根据 id 查询 user。

```
<select id="getUserList" resultType="User" >
   select * from t_user
</select>
<select id="getUser" parameterType="int"
    resultMap="userResult">
   select * from t_user where id=#{id}
</select>
```

select 元素中除了配置 id、paramterType 等属性，还可以配置其他属性。表 6-3 是 select 可配置的属性列表。

表 6-3　select 可配置的属性列表

属　性	描　述
id	在命名空间中唯一的标识符，可以被用来引用这条语句
parameterType	将要传入这条语句的参数类的完全限定名或别名。这个属性是可选的，因为 MyBatis 可以通过 TypeHandler 推断出具体传入语句的参数，默认值为 unset
resultType	从这条语句中返回的期望类型的类的完全限定名或别名。注意，如果是集合情形，那应该是集合可以包含的类型，而不能是集合本身。使用 resultType 或 resultMap，但不能同时使用
resultMap	外部 resultMap 的命名引用。结果集的映射是 MyBatis 最强大的特性，对其能够很好地理解的话，许多复杂映射的情形都能迎刃而解。使用 resultMap 或 resultType，但不能同时使用
flushCache	将其设置为 true，任何时候只要语句被调用，都会导致本地缓存和二级缓存被清空，默认值为 false
useCache	将其设置为 true，将会导致本条语句的结果被二级缓存。默认值是对 select 元素为 true
timeout	这个设置是在抛出异常之前，驱动程序等待数据库返回请求结果的秒数。默认值为 unset（依赖驱动）
fetchSize	这是尝试影响驱动程序每次批量返回的结果行数和这个设置值相等。默认值为 unset（依赖驱动）
statementType	STATEMENT，PREPARED 或 CALLABLE 的一个。这会让 MyBatis 分别使用 Statement、PreparedStatement 或 CallableStatement，默认值为 PREPARED
resultSetType	FORWARD_ONLY，SCROLL_SENSITIVE 或 SCROLL_INSENSITIVE 中的一个，默认值为 unset （依赖驱动）
databaseId	如果配置了 databaseIdProvider，MyBatis 会加载所有不带 databaseId 或匹配当前 databaseId 的语句；如果带或者不带的语句都有，则不带的会被忽略
resultOrdered	这个设置仅针对嵌套结果 select 语句适用：如果为 true，就是假设包含了嵌套结果集或是分组了，这样的话当返回一个主结果行的时候就不会发生对前面结果集的引用情况。这就使得在获取嵌套的结果集的时候不至于导致内存不够用。默认值为 false
resultSets	这个设置仅对多结果集的情况适用，它将列出语句执行后返回的结果集并为每个结果集给一个名称，名称是逗号分隔的

6.3.2　更新元素 Insert、Update、Delete

Insert、Update、Delete 三个元素都是更新操作，实现也比较相似。表 6-4 所示是三个元素对应的属性。

表 6-4　Insert、Update、Delete 三个元素对应的属性

属　　性	描　　述
id	在命名空间中唯一的标识符，可以被用来引用这条语句
parameterType	将要传入这条语句的参数类的完全限定名或别名。这个属性是可选的，因为 MyBatis 可以通过 TypeHandler 推断出具体传入语句的参数，默认值为 unset
flushCache	将其设置为 true，任何时候只要语句被调用，都会导致本地缓存和二级缓存被清空，默认值为 false
timeout	这个设置是在抛出异常之前，驱动程序等待数据库返回请求结果的秒数，默认值为 unset（依赖驱动）
statementType	STATEMENT、PREPARED 或 CALLABLE 的一个。这会让 MyBatis 分别使用 Statement、PreparedStatement 或 CallableStatement，默认值为 PREPARED
useGeneratedKeys	仅对 insert 和 update 有用，这会令 MyBatis 使用 JDBC 的 getGeneratedKeys 方法来取出由数据库内部生成的主键（比如：像 MySQL 和 SQL Server 这样的关系数据库管理系统的自动递增字段），默认值为 false
keyProperty	仅对 insert 和 update 有用，唯一标记一个属性，MyBatis 会通过 getGeneratedKeys 的返回值或者通过 insert 语句的 selectKey 子元素设置它的键值，默认为 unset。如果希望得到多个生成的列，也可以是逗号分隔的属性名称列表
keyColumn	仅对 insert 和 update 有用，通过生成的键值设置表中的列名，这个设置仅在某些数据库（像 PostgreSQL）是必需的，当主键列不是表中的第一列的时候需要设置。如果希望得到多个生成的列，也可以是逗号分隔的属性名称列表
databaseId	如果配置了 databaseIdProvider，MyBatis 会加载所有不带 databaseId 或匹配当前 databaseId 的语句；如果带或者不带的语句都有，则不带的会被忽略

在 UserMapper.xml 中，我们使用了 insert 元素来插入数据，同时获取自增主键 id 值。

```
  <insert id="addUser" parameterType="user">
    <selectKey resultType="java.lang.Integer" order="AFTER"
keyProperty="id">
        SELECT LAST_INSERT_ID()
    </selectKey>
      insert into t_user(name,age,money,status) values (#{name},#{age},
#{money},#{status,typeHandler=com.demo.model.CusValuedEnumTypeHandler})
  </insert>
```

上面使用 selectKey 元素来获取插入数据时的自增主键。表 6-5 所示是 selectKey 元素的主要属性。

表 6-5　selectKey 元素的主要属性

属　性	描　述
id	在命名空间中唯一的标识符，可以被用来引用这条语句
parameterType	将要传入这条语句的参数类的完全限定名或别名。这个属性是可选的，因为 MyBatis 可以通过 TypeHandler 推断出具体传入语句的参数，默认值为 unset
resultType	从这条语句中返回的期望类型的类的完全限定名或别名。注意，如果是集合情形，那应该是集合可以包含的类型，而不能是集合本身。使用 resultType 或 resultMap，但不能同时使用
resultMap	外部 resultMap 的命名引用。结果集的映射是 MyBatis 最强大的特性，对其能够很好地理解的话，许多复杂映射的情形都能迎刃而解。使用 resultMap 或 resultType，但不能同时使用
flushCache	将其设置为 true，任何时候只要语句被调用，都会导致本地缓存和二级缓存被清空，默认值为 false
useCache	将其设置为 true，将会导致本条语句的结果被二级缓存，默认值是对 select 元素为 true

6.3.3　可重用语句块 sql

sql 元素可以被用来定义可重用的 SQL 代码段，可以包含在其他语句中。它可以被静态地（在加载参数时）参数化。不同的属性值通过包含的实例变化。有时 SQL 中有一些是重复的语句块，可以使用 sql 元素来实现 SQL 语句块的复用。这里以下面的 select 查询语句为例子来了解 sql 元素的使用。该 SQL 语句查询别名为 t1 的 t_user 表中的 id、name、age、money、status 字段。

```
select t1.id,t1.name,t1.age,t1.money,t1.status from t_user as t1
```

下面的 xml 配置将查询字段、表名、from 部分分别作为一个 sql 片段。sql 片段可以通过 include 被包含在其他语句中（其他语句也包含 sql 片段），属性值也可以被用在 include 元素的 refid 属性里或 include 内部语句中。userColumns、sometable 都是将属性值放在 include 内部语句中，someinclude 将属性值放在了 refid 属性中。

```
<sql id="userColumns">
  ${alias}.id,${alias}.name,${alias}.age,${alias}.money,${alias}.status
</sql>
<sql id="sometable">
    ${prefix}user as ${alias}
</sql>
<sql id="someinclude">
    from
      <include refid="${include_target}"/>
</sql>
<select id="getUserList2" resultMap="userResult">
    select
     <include refid="userColumns"><property name="alias" value="t1"/>
    </include>
    <include refid="someinclude">
      <property name="prefix" value="t_"/>
      <property name="alias" value="t1"/>
```

```
        <property name="include_target" value="sometable"/>
      </include>
</select>
```

这里为了演示 sql 元素的使用，把一条简单的 select 查询语句拆得比较零散，在实际项目中可能没必要拆得那么复杂。不过适当的拆分还是有必要的，这样也能复用一些 sql 片段。

6.3.4　数据集映射 resultMap

细心的话可能会从前面章节中发现 select 查询用户标签有两种写法：一个是使用 resultType 来设置返回类型，一个是使用 resultMap 来设置返回类型。

```
<select id="getUserList" resultType="User">
    select * from t_user
</select>
<select id="getUserList" resultMap="userResult">
    select * from t_user
</select>
```

这两种写法都是正确的，那么 resultType 和 resultMap 有什么关系呢？在 MyBatis 进行查询映射的时候，其实查询出来的每一个属性都是放在一个对应的 Map 里面的，其中键是属性名、值是对应的值。当提供的返回类型属性是 resultType 的时候，MyBatis 会将 Map 里面的键值对取出赋给 resultType 所指定的对象对应的属性。所以其实 MyBatis 的每一个查询映射的返回类型都是 ResultMap，只是当我们提供的返回类型属性是 resultType 的时候，MyBatis 会自动地给我们把对应的值赋给 resultType 指定对象的属性，而当我们提供的返回类型是 resultMap 的时候，因为 Map 不能很好地表示领域模型，我们就需要自己再进一步地把它转化为对应的对象，这常常在复杂查询中很有作用。下面是 resultMap 元素的概念视图。

resultMap

- constructor：用于在实例化类时注入结果到构造方法中。
  - idArg：ID 参数，标记出作为 ID 的结果可以帮助提高整体性能。
  - arg：将被注入到构造方法的一个普通结果。
- id：一个 ID 结果，标记出作为 ID 的结果可以帮助提高整体性能。
- result：注入到字段或 JavaBean 属性的普通结果。
- association：一个复杂类型的关联，许多结果将包装成这种类型。
  - 嵌套结果映射：关联可以指定为一个 resultMap 元素，或者引用一个。
- collection：一个复杂类型的集合。
  - 嵌套结果映射：集合可以指定为一个 resultMap 元素，或者引用一个。
- discriminator：使用结果值来决定使用哪个 resultMap。
  - case：基于某些值的结果映射。
    - 嵌套结果映射：一个 case 也是一个映射它本身的结果，因此可以包含很多相同的元素，或者参照一个外部的 resultMap。

通常一个复杂数据类型在 XML 中的配置格式如下:

```
<!--column 不做限制，可以为任意表的字段，而 property 须为 type 定义的 pojo 属性-->
<resultMap id="唯一的标识" type="映射的 pojo 对象">
  <id column="表的主键字段，或者可以为查询语句中的别名字段" jdbcType="字段类型"
property="映射 pojo 对象的主键属性" />
  <result column="表的一个字段（可以为任意表的一个字段）" jdbcType="字段类型"
property="映射到 pojo 对象的一个属性（须为 type 定义的 pojo 对象中的一个属性）"/>
  <association property="pojo 的一个对象属性" javaType="pojo 关联的 pojo 对象">
    <id column="关联 pojo 对象对应表的主键字段" jdbcType="字段类型" property="关
联 pojo 对象的主键属性"/>
    <result  column="任意表的字段" jdbcType="字段类型" property="关联 pojo 对象
的属性"/>
  </association>
  <!-- 集合中的 property 须为 oftype 定义的 pojo 对象的属性-->
  <collection property="pojo 的集合属性" ofType="集合中的 pojo 对象">
    <id column="集合中 pojo 对象对应的表的主键字段" jdbcType="字段类型" property="
集合中 pojo 对象的主键属性" />
    <result column="可以为任意表的字段" jdbcType="字段类型" property="集合中的
pojo 对象的属性" />
  </collection>
</resultMap>
```

表与表之间的数据映射主要有一对一、一对多和多对多的关系。这里也主要介绍 association、collection 的使用。为了演示，准备了 4 张数据表 card、course、role 和 user_role 以及 3 张表对应的 model。t_user 与 card 是一对一关系，t_user 与 course 是一对多关系，t_user 与 role 是多对多关系。为了保存多对多关系，增加了 user_role 表。

card 表：

```
CREATE TABLE `card` (
  `id` int(11) NOT NULL AUTO_INCREMENT,
  `cardNo` varchar(20) NOT NULL,
  `city` varchar(45) NOT NULL,
  `address` varchar(100) NOT NULL,
  `userid` int(11) NOT NULL,
  PRIMARY KEY (`id`)
) ENGINE=InnoDB AUTO_INCREMENT=2 DEFAULT CHARSET=utf8;
```

card 类：

```
public class Card {

    @Override
    public String toString() {
        return  "Card [id=" + getId() + ", cardNo=" + cardNo + ", city=" + city
+ ",address="+address+",userid="+userid+"]";
    }
```

```
    private int id;
    private String cardNo;
    private String city;
    private String address;
    private int userid;
    public int getUserid() {
        return userid;
    }
    public void setUserid(int userid) {
        this.userid = userid;
    }
    public String getAddress() {
        return address;
    }
    public void setAddress(String address) {
        this.address = address;
    }
    public int getId() {
        return id;
    }
    public void setId(int id) {
        this.id = id;
    }
    public String getCardNo() {
        return cardNo;
    }
    public void setCardNo(String cardNo) {
        this.cardNo = cardNo;
    }
    public String getCity() {
        return city;
    }
    public void setCity(String city) {
        this.city = city;
    }
}
```

course 表：

```
CREATE TABLE `course` (
  `id` int(11) NOT NULL AUTO_INCREMENT,
  `name` varchar(45) NOT NULL,
  `userid` int(11) NOT NULL,
  PRIMARY KEY (`id`)
) ENGINE=InnoDB AUTO_INCREMENT=3 DEFAULT CHARSET=utf8;
```

course 类：

```
public class Course {
    @Override
    public String toString() {
        return  "Course [id=" + getId() + ", name=" + name +",userid=
"+userid+"]";
    }
    private int id;
    private String name;
    private int userid;
    public int getUserid() {
        return userid;
    }
    public void setUserid(int userid) {
        this.userid = userid;
    }
    public int getId() {
        return id;
    }
    public void setId(int id) {
        this.id = id;
    }
    public String getName() {
        return name;
    }
    public void setName(String name) {
        this.name = name;
    }
}
```

role 表：

```
CREATE TABLE `role` (
  `id` int(11) NOT NULL AUTO_INCREMENT,
  `name` varchar(45) NOT NULL,
  `desp` varchar(45) NOT NULL,
  PRIMARY KEY (`id`)
) ENGINE=InnoDB AUTO_INCREMENT=3 DEFAULT CHARSET=utf8;
```

role 类：

```
public class Role {
    @Override
    public String toString() {
        // TODO Auto-generated method stub
        return  "Role [id=" + getId() + ", name=" + name +",desp="+desp+"]";
```

```
    }
    private int id;
    private String name;
    private String desp;
    private List<User> users;
    public int getId() {
        return id;
    }
    public void setId(int id) {
        this.id = id;
    }
    public String getName() {
        return name;
    }
    public void setName(String name) {
        this.name = name;
    }
    public String getDesp() {
        return desp;
    }
    public void setDesp(String desp) {
        this.desp = desp;
    }
    public List<User> getUsers() {
        return users;
    }
    public void setUsers(List<User> users) {
        this.users = users;
    }

}
```

user_role 表：

```
CREATE TABLE `user_role` (
  `userid` int(11) NOT NULL,
  `roleid` int(11) NOT NULL,
  PRIMARY KEY (`userid`,`roleid`)
) ENGINE=InnoDB DEFAULT CHARSET=utf8;
```

上面把数据表的关系和单个实体类已经准备好，下面就需要准备实体类之间的关系，所以需要在 user 类中增加下面几个属性来表示实体类之间的关系。

```
    private Card card;
    public Card getCard() {
        return card;
    }
    public void setCard(Card card) {
```

```
        this.card = card;
    }
    private List<Role> roles;
    private List<Course> courses;

    public List<Course> getCourses() {
        return courses;
    }
    public void setCourses(List<Course> courses) {
        this.courses = courses;
    }
    public List<Role> getRoles() {
        return roles;
    }
    public void setRoles(List<Role> roles) {
        this.roles = roles;
    }
```

现在数据表和实体类都已准备完成。下面就是将一个复杂查询的结果映射到实体类上。

```
    <select id="getuser" parameterType="int"
         resultMap="userResultTest">
     select a.id as user_id,a.name as user_name,a.age as user_age,a.money as
user_money,a.status as user_status,b.id as card_id,b.cardNo as
card_cardNo,b.userid as card_userid,b.city as card_city,b.address as
card_address,c.id as course_id,c.name as course_name,c.userid as
course_userid,e.name as role_name,e.desp as role_desp from t_user a left join
card b on a.id=b.userid left join course c on a.id=c.userid left join user_role
d on a.id=d.userid left join role e on d.roleid=e.id where a.id=#{id}
     </select>
```

为了使用演示表之间一对一、一对多和多对多的关系，上面的 sql 通过 t_user 依次左连接 card、course、user_role 和 role 表。而该 sql 对应的 xml 映射如下：

```
    <resultMap type="User" id="userResultTest">
        <result column="user_id" property="id"/>
        <result column="user_name" property="name"/>
        <result column="user_age" property="age"/>
  <result column="user_money" property="money"/>
        <result column="user_status" property="status" typeHandler=
"com.demo.model.CusEnumStatusHandler"/>
        <association property="card" javaType="Card" columnPrefix="card_">
            <result column="id" property="id"/>
            <result column="no" property="cardNo"/>
            <result column="city" property="city"/>
            <result column="address" property="address"/>
            <result column="userid" property="userid"/>
        </association>
```

```
            <collection property="courses" javaType="ArrayList" ofType="Course"
columnPrefix="course_">
                <result column="id" property="id"/>
                <result column="name" property="name"/>
                <result column="userid" property="userid"/>
            </collection>
            <collection property="roles" javaType="ArrayList" ofType="Role"
columnPrefix="role_">
                <result column="id" property="id"/>
                <result column="name" property="name"/>
                <result column="desp" property="desp"/>
                <result column="userid" property="userid"/>
            </collection>
        </resultMap>
```

在 resultMap 中 user 与 card 对象是一对一关系，使用 association 关联，对于一对多的关系使用 collection 关联，对于多对多的关系可以理解为两个一对多的关系。现在又出现一个新的问题，在映射 Card、Course 和 Role 的时候把映射关系都放在 id=userResult 的 resultMap 中，这样如果以后要映射 Card、Course、Role 的时候还要写一遍，复用性不高。其实我们可以在 association 和 collection 节点增加属性 resultMap。

```
        <resultMap type="Card" id="cardResult">
            <result column="id" property="id"/>
            <result column="no" property="cardNo"/>
            <result column="city" property="city"/>
            <result column="address" property="address"/>
            <result column="userid" property="userid"/>
        </resultMap>

        <resultMap type="Course" id="courseResult">
            <result column="id" property="id"/>
            <result column="name" property="name"/>
            <result column="userid" property="userid"/>
        </resultMap>
        <resultMap type="Role" id="roleResult">
            <result column="id" property="id"/>
            <result column="name" property="name"/>
            <result column="desp" property="desp"/>
            <result column="userid" property="userid"/>
        </resultMap>
        <resultMap type="User" id="userResultTest1">
            <result column="user_id" property="id"/>
            <result column="user_name" property="name"/>
            <result column="user_age" property="age"/>
    <result column="user_money" property="money"/>
            <result column="user_status" property="status" typeHandler=
"com.demo.model.CusEnumStatusHandler"/>
```

```
            <association property="card" javaType="Card" resultMap="cardResult"
 columnPrefix="card_"></association>
            <collection property="courses" javaType="ArrayList" ofType="Course"
 resultMap="courseResult" columnPrefix="course_"></collection>
            <collection property="roles" javaType="ArrayList" ofType="Role"
 resultMap="roleResult" columnPrefix="role_"></collection>
        </resultMap>
```

这里先实现了单个表的映射关系，然后在 association 和 collection 中使用 resultMap 引用这些单个表的 resultMap，提高了复用性。

在 resultMap 元素中还有一个重要的子元素 discriminator（鉴别器）。有时一个单独的数据库查询也会返回很多不同（但是希望有些关联）数据类型的结果集，或者有时一个父类有多个子类，每个子类有不同的属性，而每个基类对应的数据库表是同一个表，只是部分字段为 null。鉴别器元素就是被设计来处理这些情况的。鉴别器非常容易理解，因为它的表现很像 Java 语言中的 switch 语句。

```
        <resultMap type="user" id="userResultTest2">
            <id column="user_id" property="id"/>
            <result column="user_name" property="name"/>
            <result column="user_status" property="status" typeHandler=
 "com.demo.model.CusEnumStatusHandler"/>
            <association property="card" javaType="Card" resultMap="cardResult"
 columnPrefix="card_"></association>
            <collection property="courses" javaType="ArrayList" ofType="Course"
 resultMap="courseResult" columnPrefix="course_"></collection>
            <collection property="roles" javaType="ArrayList" ofType="Role"
 resultMap="roleResult" columnPrefix="role_"></collection>
            <discriminator javaType="string" column="user_status">
            <case value="0" resultType="User">
                <result column="user_age" property="age"/>
            </case>
            <case value="1" resultType="User">
                 <result column="user_money" property="money"/>
            </case>
            </discriminator>
        </resultMap>
```

在上面的 resultMap 中，根据 user_status 判断，若为 1，则映射 age 属性忽略 money 属性，若为 1，则映射 money 属性忽略 age 属性，这里的 resultType 都是 User，如果是多个子类的情况，可以设置 resultType 为不同的子类。

6.3.5 缓存和自定义缓存

MyBatis 包含一个非常强大的查询缓存特性，可以非常方便地配置和定制。本节来学习一下 MyBatis 缓存和自定义缓存。

1. 自带缓存

MyBatis 默认开启了一级缓存，一级缓存是在 SqlSession 层面进行缓存的，即同一个 SqlSession，多次调用同一个 Mapper 和同一个方法的同一个参数，只会进行一次数据库查询，然后把数据缓存到缓冲中，以后直接从缓存中取出数据，不会直接去查数据库。不同的 SqlSession 对象，因为不用的 SqlSession 都是相互隔离的，所以相同的 Mapper、参数和方法还是会再次发送 SQL 到数据库去执行返回结果。

为了克服这个问题，需要开启二级缓存，在 SqlSessionFactory 层面给各个 SqlSession 对象共享。默认二级缓存是不开启的，需要手动进行配置。手动开启二级缓存比较简单，首先开启缓存总开关，在 MyBatis 的配置文件 settings 元素设置 cacheEnabled 参数为 True，然后只需在 SQL 映射文件中添加一行：<cache/>。如果这样配置的话，很多其他的配置就会被默认进行，例如：

（1）映射文件所有的 select 语句会被缓存。

（2）映射文件的所有 insert、update 和 delete 语句会刷新缓存。

（3）缓存会使用默认的 Least Recently Used（LRU，最近最少使用原则）的算法来回收缓存空间。

（4）根据时间表，比如 No Flush Interval（CNFI，没有刷新间隔），缓存不会以任何时间顺序来刷新。

（5）缓存会存储列表集合或对象（无论查询方法返回什么）的 1024 个引用。

（6）缓存会被视为是 read/write（可读/可写）的缓存，意味着对象检索不是共享的，而且可以很安全地被调用者修改，不干扰其他调用者或线程所做的潜在修改。

上面是使用的默认属性设置，当然也可以手动配置一些属性。

```
<cache eviction="LRU" flushInterval="100000" size="1024" readOnly="true"/>
```

各个属性的意义如下：

- eviction：缓存回收策略。
  - LRU：最少使用原则，移除最长时间不使用的对象。
  - FIFO：先进先出原则，按照对象进入缓存顺序进行回收。
  - SOFT：软引用，移除基于垃圾回收器状态和软引用规则的对象。
  - WEAK：弱引用，更积极地移除基于垃圾回收器状态和弱引用规则的对象。
- flushInterval：刷新时间间隔，单位为毫秒，这里配置的是 100 秒。如果不配置，那么只有在进行数据库修改操作时才会被动刷新缓存区。
- size：引用额数目，代表缓存最多可以存储的对象个数。
- readOnly：是否只读，如果为 true，则所有相同的 sql 语句返回的是同一个对象（有助于提高性能，但并发操作同一条数据时，可能不安全）；如果设置为 false，则相同的 sql 后面访问的是 cache 的 clone 副本。

如果开启了二级缓存，映射文件所有的 select 语句就会被缓存而且映射文件的所有 Insert、

Update 和 Delete 语句会刷新缓存，但有时也会遇到不需要进行缓存操作实时性较高的 select 语句，或者执行 Insert、Update 和 Delete 语句之后不需要刷新缓存的。针对这些情况，可以在 Mapper 的具体方法下设置对二级缓存的访问意愿。

如果一条语句每次都需要最新的数据，就意味着每次都需要从数据库中查询数据，可以把 useCache 这个属性设置为 false，例如：

```
<select id="getUserList" resultType="User" useCache="false">
   select * from t_user
</select>
```

二级缓存默认会在 Insert、Update、Delete 操作后刷新缓存，可以使用 flushCache 属性设置为 false，手动配置不更新缓存，例如：

```
<insert id="addUser" parameterType="user" flushCache="false">
```

2. 自定义缓存

缓存的介质可以有多种。MyBatis 提供了 org.apache.ibatis.cache.Cache 接口，用来自定义缓存。这里新建一个 CusCache 类，实现 Cache 接口。在 CusCache 类中定义一个 ConcurrentHashMap<Object,Object>类型的 cache 变量来保存缓存。

```
package com.demo.model;
import java.util.concurrent.ConcurrentHashMap;
import java.util.concurrent.locks.ReadWriteLock;
import java.util.concurrent.locks.ReentrantReadWriteLock;
import org.apache.ibatis.cache.Cache;

public class CusCache implements Cache {

    private ReadWriteLock lock = new ReentrantReadWriteLock();
    private ConcurrentHashMap<Object,Object> cache = new ConcurrentHashMap
<Object, Object>();
    private String id;

    public  CusCache(){
        System.out.println("初始化-1! ");
    }

   //必须有该构造函数
    public CusCache(String id){
        System.out.println("初始化-2! ");
        this.id = id;
    }

    // 获取缓存编号
    public String getId()  {
        System.out.println("得到 ID: " + id);
```

```
        return id;
    }

    //获取缓存对象的大小
    public int getSize() {
        System.out.println("获取缓存大小！");
        return 0;
    }
    // 保存 key 值缓存对象
    public void putObject(Object key, Object value) {
        System.out.println("往缓存中添加元素：key=" + key+",value=" + value);
        cache.put(key,value);
    }

    //通过 KEY
    public Object getObject(Object key) {
        System.out.println("通过 kEY 获取值：" + key);
        System.out.println("OVER");
        System.out.println ("===============================================");
        System.out.println("值为：" + cache.get(key));
        System.out.println ("=====================OVER==================");
        return cache.get(key);
    }

    // 通过 key 删除缓存对象
    public Object removeObject(Object key) {
        System.out.println("移除缓存对象：" + key);
        return null;
    }

    // 清空缓存
    public void clear() {
        System.out.println("清除缓存！");
        cache.clear();
    }

    // 获取缓存的读写锁
    public ReadWriteLock getReadWriteLock() {
        System.out.println("获取锁对象！！！");
        return lock;
    }
}
```

自定义缓存之后可以进行验证。在 main 方法中，实例化了两个 SqlSession 对象，在第一个 SqlSession 中对同一个 statement 执行两次 select 查询再 commit，第二个 SqlSession 中对同一个 statement 执行同样的查询并 commit。

```
            String resource = "mybatis-config.xml";
            //使用 MyBatis 提供的 Resources 类加载 mybatis 的配置文件（它也加载关联的映射
文件）
            Reader reader = Resources.getResourceAsReader(resource);
            //构建 sqlSession 的工厂
            SqlSessionFactory sessionFactory = new SqlSessionFactoryBuilder().
build(reader);
            //创建能执行映射文件中 sql 的 sqlSession
            SqlSession session = sessionFactory.openSession();
            String statement="com.demo.mybatis.DBMapping.UserMapper.getuser";
            User user=session.selectOne(statement, 1);
            user=session.selectOne(statement, 1);
            System.out.println(user.toString());
            session.commit();
            SqlSession session1 = sessionFactory.openSession();
            user=session1.selectOne(statement, 1);
            System.out.println(user.toString());
            session1.commit();
```

运行上面的代码，日志部分截图如图 6-1 所示。

```
OVER
========================================================
值为: null
====================OVER==============================
Id:1 Name:张三Age:0 Money:377.77 Status:AVAILABLE
在缓存中添加元素: key=-731237125:4770985981:com.demo.mybatis.DBMapping.UserMapper.getuser:0:214748
user_age,a.money as user_money,a.status as user_status,b.id as card_id,b.cardNo as card_car
      b.city as card_city,b.address as card_address,c.id as course_id,c.name as course_na
role_desp
      from t_user a left join card b on a.id=b.userid left join course c on a.id=c.userid
on d.roleid=e.id
      where a.id=?:1:development,value=[Id:1 Name:张三Age:0 Money:377.77 Status:AVAILABLE]
通过KEY获取值: -731237125:4770985981:com.demo.mybatis.DBMapping.UserMapper.getuser:0:2147483647
user_age,a.money as user_money,a.status as user_status,b.id as card_id,b.cardNo as card_car
      b.city as card_city,b.address as card_address,c.id as course_id,c.name as course_na
role_desp
      from t_user a left join card b on a.id=b.userid left join course c on a.id=c.userid
on d.roleid=e.id
      where a.id=?:1:development
OVER
========================================================
值为: [Id:1 Name:张三Age:0 Money:377.77 Status:AVAILABLE]
====================OVER==============================
Id:1 Name:张三Age:0 Money:377.77 Status:AVAILABLE
```

图 6-1

从日志中可以看出，每次查询数据库前，MyBatis 都会先在缓存中查找是否有该缓存对象。只有当调用了 commit()方法，MyBatis 才会往缓存中写入数据，数据记录的键为“数字编号+Mapper 名+方法名+SQL 语句+参数”格式，值为返回的对象值。所以在第一个 SqlSession 对象 commit()之前，虽然 key 相同，但通过 key 获取 cache 缓存对象都为 null，commit()之后再通过该 key 获取缓存时值是存在的。

6.4 动态 SQL

MyBatis 还有一个方便的功能就是动态 SQL，可以根据条件智能生成 SQL 语句。MyBatis 通过 if、choose、when、otherwise、trim、where、set、foreach 等标签，可组合成非常灵活的 SQL 语句，从而在提高 SQL 语句的准确性的同时，也大大提高了开发人员的效率。

6.4.1 if 语句

有些页面提供多条件查询功能，输入值时进行模糊查询，不输入值时不做过滤条件。这样的话就需要拼接 where，一般在 sql 后面加 where 1=1，可以不为 where 和 and 揪心。

```
<select id="finduserbylikename1" parameterType="string" resultMap=
"userResult">
    select * from t_user where 1=1
    <if test="_parameter!=null and _parameter!=''">
        and name like #{_parameter}
    </if>
</select>
<select id="finduserbylikename2" parameterType="map" resultMap=
"userResult">
  <bind name="pattern" value="'%' + _parameter.name + '%'" />
    select * from t_user where 1=1
    <if test="_parameter.name!=null and _parameter.name!=''">
        and name like #{pattern}
    </if>
</select>
```

上面用了两种传参数的方式：一种是 string，另一种是 map。用 map 可以传多个参数。还有就是在第二种中使用了 bind，可以在 sql 中增加%%。参数中可以不带，下面是两个传参的使用。

```
String statement="com.demo.mybatis.DBMapping.UserMapper.
finduserbylikename1";
List<User> users=session.selectList(statement, "%J%");
System.out.println(users.size());
for(int i=0;i<users.size();i++)
{
    System.out.println(users.get(i).toString());
}
statement="com.demo.mybatis.DBMapping.UserMapper.
finduserbylikename2";
Map<String,Object> map=new HashMap<String,Object>();
map.put("name", "J");
```

```
users=session.selectList(statement, map);
System.out.println(users.size());
for(int i=0;i<users.size();i++)
{
    System.out.println(users.get(i).toString());
}
```

6.4.2 choose (when, otherwise)语句

做项目也经常会遇到左边一个下拉框选择搜索类型，按什么类型搜索，右边一个文本框，输入搜索关键字，点查询搜索。如果遇到这种情况，用 choose 再合适不过了。

```
<select id="findcard" parameterType="map" resultMap="cardResult">
 <bind name="pattern" value="'%' + _parameter.value + '%'" />
   select * from card where 1=1
     <choose>
          <when test="type=='city'">
            AND city like #{pattern}
          </when>
          <when test="type=='address'">
            AND address like #{pattern}
          </when>
          <otherwise>
            AND city like #{pattern} or address like #{pattern}
          </otherwise>
     </choose>
</select>
```

这里根据 type 来判断过滤哪些字段。如果是 city，就使用关键字过滤 city 字段；如果是 address，就使用关键字过滤 address 字段；如果是其他，就为 city 和 address 过滤的并集。

6.4.3 choose (when, otherwise)语句

对对象的修改操作，可能有时需要根据不同的条件修改不同的列。比如根据 id 修改 name，根据 id 修改 age，每个都写一个 sql，这样也太不灵活了，于是 trim 出现了。trim 标记是一个格式化的标记，可以完成 set 或者 where 标记的功能。prefix:在 trim 标签内为 sql 语句加上前缀，suffix:在 trim 标签内为 sql 语句加上后缀，suffixOverrides:指定去除多余的后缀内容。例如："suffixOverrides=","" 去除 trim 标签内 sql 语句多余的后缀","，"prefixOverrides:" 指定去除多余的前缀内容。

```
<update id="updatecard" parameterType="map">
 update card
 <trim prefix="set" suffixOverrides=",">
       <if test="_parameter.cardNo!=null and _parameter.cardNo!=''">
             cardNo = #{cardNo},
     </if>
     <if test=" _parameter.city!=null and _parameter.city!=''">
```

```
                city = #{city},
            </if>
        </trim>
        <trim prefix="WHERE" prefixOverrides="and|or">
            <if test=" _parameter.id!=null and _parameter.id!=''">
                and id = #{id}
            </if>
            <if test=" _parameter.address!=null and _parameter.address!=''">
                and address = #{address}
            </if>
        </trim>
    </update>
```

有了 trim，可以很方便地写出 where、set 子句，而不用担心 set 子句的逗号、where 子句的 and 和 or。

6.4.4　foreach 语句

有时候 sql 需要使用 in 查询，通常传的是一个 collection 集合，那该怎么样拼接呢？当然是 foreach。foreach 元素主要有下面几个属性：

- Item：表示集合中每一个元素进行迭代时的别名。
- Index：指定一个名字，用于表示在迭代过程中每次迭代到的位置。
- Open：表示该语句以什么开始。
- Separator：表示在每次进行迭代之间以什么符号作为分隔符。
- Close：表示以什么结束。

在下面的例子中，select 查询根据 id 查询用户，parameterType 为 map。

```
    <select id="findcardbyuserids" parameterType="map" resultMap=
"userResult">
        select * from t_user where id in
        <foreach collection="ids" index="index" item="item" open="("
separator="," close=")">
            #{item}
        </foreach>
    </select>
```

在 main 中调用时需要为该查询语句传递一个 map 参数。

```
        String statement="com.demo.mybatis.DBMapping.UserMapper.
findcardbyuserids";
        List ids = new ArrayList();
        ids.add("1");
        ids.add("2");
        ids.add("3");
        Map<String,Object> map=new HashMap<String,Object>();
        map.put("ids", ids);
```

```
List<User> users=session.selectList(statement,map);
for(int i=0;i<users.size();i++)
{
   System.out.println(users.get(i).toString());
}
```

6.5 逆向工程

MyBatis 的一个主要特点就是需要程序员自己编写 sql，如果表太多的话，难免会很麻烦，所以 MyBatis 官方提供了一个逆向工程，可以针对单表自动生成 MyBatis 执行所需要的代码（包括 mapper.xml、mapper.java）。一般在开发中，常用的逆向工程方式是通过数据库的表生成代码。

要使用 generator 插件自动生成相关文件，需要引入 mybatis-generator-core 这个包，在 <dependencys>中加入：

```
<dependency>
    <groupId>org.mybatis.generator</groupId>
    <artifactId>mybatis-generator-core</artifactId>
    <version>1.3.2</version>
</dependency>
```

引入依赖之后需要引入插件 maven-compiler-plugin 和 mybatis-generator-maven-plugin，同时设置自动生成代码配置：

```
<plugin>
    <groupId>org.apache.maven.plugins</groupId>
    <artifactId>maven-compiler-plugin</artifactId>
    <configuration>
       <source>1.8</source>
       <target>1.8</target>
    </configuration>
    <version>3.3</version>
</plugin>
<plugin>
    <groupId>org.mybatis.generator</groupId>
    <artifactId>mybatis-generator-maven-plugin</artifactId>
    <version>1.3.2</version>
   <dependencies>
       <dependency>
          <groupId>mysql</groupId>
          <artifactId>mysql-connector-java</artifactId>
          <version>5.1.35</version>
       </dependency>
    </dependencies>
    <configuration>
```

```
            <!--配置文件的路径-->
            <configurationFile>src/main/resources/generatorConfig.xml
</configurationFile>
            <overwrite>true</overwrite>
        </configuration>
    </plugin>
```

在引入插件时需要设置自动生成插件的配置文件，这里新建一个 generatorConfig.xml 文件，在文件中以 t_user 表为例来了解自动生成插件的使用。

```
<?xml version="1.0" encoding="UTF-8"?>
<!DOCTYPE generatorConfiguration
        PUBLIC "-//mybatis.org//DTD Mybatis Generator Configuration 1.0//EN"
        "http://mybatis.org/dtd/mybatis-generator-config_1_0.dtd">
<generatorConfiguration>
   <context id="DB2Tables"   targetRuntime="MyBatis3">
      <commentGenerator>
         <property name="suppressDate" value="true"/>
         <property name="suppressAllComments" value="true"/>
      </commentGenerator>
      <!--数据库驱动，数据库地址及表名，账号，密码-->
      <jdbcConnection driverClass="com.mysql.jdbc.Driver"
connectionURL="jdbc:mysql://127.0.0.1:3306/daodemodb"   userId="root"
password="123456">
      </jdbcConnection>
      <javaTypeResolver>
         <property name="forceBigDecimals" value="false"/>
      </javaTypeResolver>
      <!--生成 Model 类的包名及存放位置-->
      <javaModelGenerator targetPackage="com.demo.model"
targetProject="src/main/java">
         <property name="enableSubPackages" value="true"/>
         <property name="trimStrings" value="true"/>
      </javaModelGenerator>
      <!--生成映射文件的包名及存放位置-->
      <sqlMapGenerator targetPackage="com.demo.mybatis.DBMapping"
targetProject="src/main/java">
         <property name="enableSubPackages" value="true"/>
      </sqlMapGenerator>
      <!--生成 Dao 类的包名及存放位置-->
      <javaClientGenerator type="XMLMAPPER"
targetPackage="com.demo.model" targetProject="src/main/java">
         <property name="enableSubPackages" value="true"/>
      </javaClientGenerator>
      <!--生成对应表及类名,domainObjectName 是设置实体类的名字的-->
      <table tableName="t_user" domainObjectName="TUser"
enableCountByExample="false" enableUpdateByExample="false"
```

```
enableDeleteByExample="false" enableSelectByExample="false"
selectByExampleQueryId="false"></table>
        </context>
    </generatorConfiguration>
```

在上面的 xml 中设置了生成映射文件、DAO 类、Model 类的包名及存放位置，并设置要生成的对应的表名类名，这里只设置了 t_user 来作为参考。至此，就已配置完成，只需右击，选择 Run As→Maven build。如图 6-2 所示，在 Goals 中输入“mybatis-generator:generate”。

图 6-2

单击 Run 按钮，如果成功，就会在命令行窗口输出 BUILD SUCCESS。刷新项目，可以看到新增的文件，如图 6-3 所示。

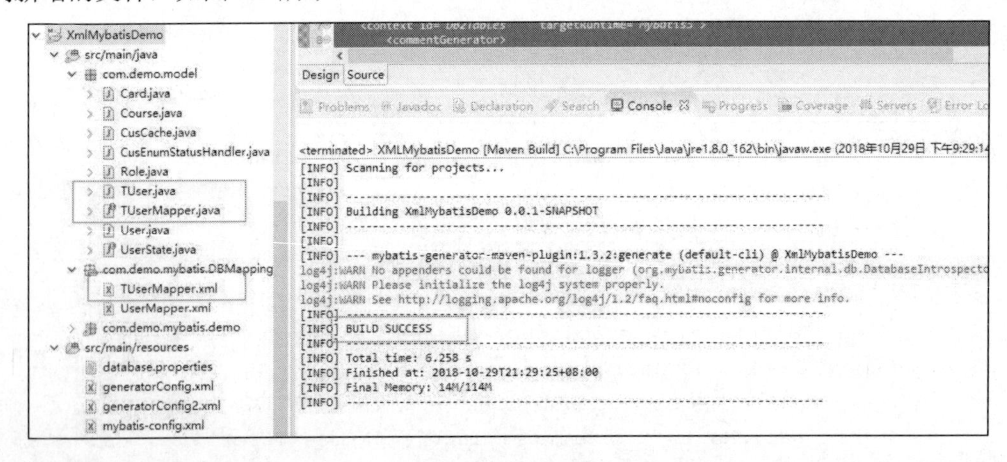

图 6-3

6.6 分页插件 pagehelper

在编写 Web 后台代码时，分页是必不可少的，当然最通常的思路是在 SQL 中使用分页关键字来进行分页。在实际开发中，更多的是使用分页插件来减少代码冗杂，使编码更加清晰。pagehelper 是 MyBatis 提供的分页插件，目前支持 Oracle、MySQL、MariaDB、SQLite、Hsqldb、PostgreSQL 六种数据库。本节主要介绍 pagehelper 的简单使用。

1. 引入依赖

把 pagehelper 所需用到的 jar 包添加到工程中，官方提供的代码对逆向工程支持不好，这里使用的是开源的 pagehelper，只需在 pom.xml 中引入 pagehelper 依赖。

```
<dependency>
   <groupId>com.github.pagehelper</groupId>
   <artifactId>pagehelper</artifactId>
   <version>3.4.2</version>
</dependency>
```

2. 配置插件

引入依赖之后，还需要在 mybatis-config.xml 中配置 pagehelper 插件，从名字也能知道是什么，相当于配置了一个拦截器。

```
<plugins>
    <!-- com.github.pagehelper 为 PageHelper 类所在包名 -->
    <plugin interceptor="com.github.pagehelper.PageHelper">
       <!-- 方言 -->
       <property name="dialect" value="mysql"/>
       <!-- 该参数默认为 false -->
       <!-- 设置为 true 时，使用 RowBounds 分页会进行 count 查询 -->
       <property name="rowBoundsWithCount" value="true"/>
    </plugin>
</plugins>
```

3. 分页测试

pagehelper 插件使用非常简单，只需上面两步即可进行分页操作，这里以查询 t_user 表为例。将 t_user 表中的数据以每页 5 行进行分页，并打印出数据总数、总页数、最后一页行数以及第一页的数据。

```
SqlSession session = sessionFactory.openSession();
String statement="com.demo.mybatis.DBMapping.UserMapper.getUserList";
PageHelper.startPage(1, 5, true);
List<User> users=session.selectList(statement);
PageInfo<User> pageInfo = new PageInfo<User>(users);
```

```
System.out.println("数据总数：" + pageInfo.getTotal());
System.out.println("数据总页数：" + pageInfo.getPages());
System.out.println("最后一页：" + pageInfo.getLastPage());
for (User u: users)
{
   System.out.println(u.toString());
}
```

输出结果如下：

```
数据总数:11
数据总页数:3
最后一页:3
Id:1 Name:JdbcDaoSupportTest Age:26 Money:333.33 Status:AVAILABLE
Id:2 Name:张三 Age:24 Money:666.66 Status:AVAILABLE
Id:3 Name:李四 Age:25 Money:888.88 Status:AVAILABLE
Id:4 Name:王二 Age:26 Money:999.99 Status:AVAILABLE
Id:5 Name:小明 Age:27 Money:555.55 Status:AVAILABLE
```

6.7 小结

本章主要介绍 MyBatis 的相关知识。通过本章的学习，大家对 ORM 框架有了进一步的了解，学习了 MyBatis 的配置文件、映射文件、动态 SQL 等内容，可以掌握 ORM 的编程思想，还有助于其他 ORM 框架的学习。

第 7 章

◂ Spring之Spring MVC ▸

JavaEE 体系结构包括 4 层，从上到下分别是应用层、Web 层、业务层、持久层。Spring 是业务层的框架。MyBatis 是持久层的框架。本章将学习 Web 层的框架——Spring MVC。

本章主要涉及到的知识点有：

- MVC 设计模式：了解 MVC 设计模式。
- Spring MVC 处理流程：了解 Spring 的处理流程、HandlerMapping 的使用。
- 数据传递：View 和 Controller 之间的数据传递。
- 拦截器：了解 Spring MVC 拦截器的使用。

7.1 MVC 框架

MVC 框架是 Web 开发的常用框架，不仅 Java 语言有 MVC 框架，C#、Python 等其他语言也有类似的 MVC 框架，那 MVC 框架是什么、由哪些部分组成、为什么会有 MVC 框架呢？

7.1.1 MVC 模式简介

MVC（Model View Controller）是模型（Model）—视图（View）—控制器（Controller）的缩写，一种软件设计典范，用一种业务逻辑、数据、界面显示分离的方法组织代码，将业务逻辑聚集到一个部件里面，在改进和个性化定制界面及用户交互的同时，不需要重新编写业务逻辑。MVC 开始是存在于桌面程序中的，M 是指业务模型，V 是指用户界面，C 则是控制器，使用 MVC 的目的是将 M 和 V 的实现代码分离，从而使同一个程序可以使用不同的表现形式，比如一批统计数据可以分别用柱状图、饼图来表示。

MVC 模式把用户界面交互分拆到不同的三种角色中，使应用程序被分成三个核心部件：Model（模型）、View（视图）、Controller（控制器）。它们各自处理自己的任务：

（1）模型：模型持有所有的数据、状态和程序逻辑，独立于视图和控制器。

（2）视图：用来呈现模型。视图通常直接从模型中取得它需要显示的状态与数据。对于相同的信息，可以有多个不同的显示形式或视图。

（3）控制器：位于视图和模型中间，负责接受用户的输入，将输入进行解析并反馈给模型，通常一个视图具有一个控制器。

模型（Model）、视图（View）、控制器（Controller）三者的关系如图 7-1 所示。

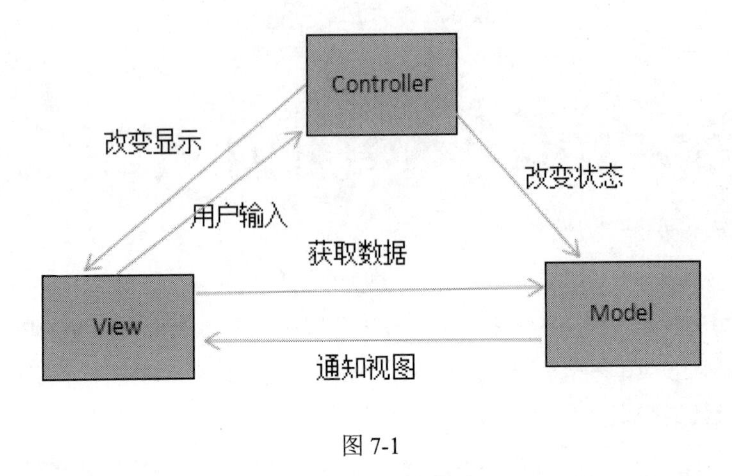

图 7-1

视图中用户的输入被控制器解析后，控制器改变状态激活模型，模型根据业务逻辑维护数据，并通知视图数据发生变化，视图得到通知后从模型中获取数据刷新自己。

7.1.2 MVC 和设计模式区别

大家往往把框架模式和设计模式混淆，认为 MVC 是一种设计模式。实际上它们完全是不同的概念。MVC 是一种使用 MVC（Model View Controller，模型-视图-控制器）设计创建 Web 应用程序的模式。它是框架的一种模式，类似 ORM，同时该模式下可能有多种框架。本章介绍的 Spring MVC 是一个框架，而这个框架的模式是 MVC 模式。设计模式是对在某种环境中反复出现的问题以及解决该问题的方案的描述，比框架更抽象；框架可以用代码表示，也能直接执行或复用，对模式而言只有实例才能用代码表示；设计模式是比框架更小的元素，一个框架中往往含有一个或多个设计模式，框架总是针对某一特定应用领域，但同一模式却可适用于各种应用。框架是软件，而设计模式是软件的知识。简而言之：框架是大智慧，用来对软件设计进行分工；设计模式是小技巧，对具体问题提出解决方案，以提高代码复用率、降低耦合度。

7.1.3 优缺点

MVC 的优点体现在以下几个方面：

（1）有利于团队开发分工协作和质量控制，降低开发成本。在开发过程中，可以更好地分工、更好地协作，有利于开发出高质量的软件。良好的项目架构设计，将减少编码工作量：采用 MVC 结构 + 代码生成器，是大多数 Web 应用的理想选择。部分模型（Model）和存储过程一般可用工具自动生成。控制器（Controller）比较稳定，一般由架构师（也可能是有经验的人）完成。整个项目需要手动编写代码的地方只有视图（View）。在这种模式下，个人能力不是特别重要，只要懂点语法基础的人都可以编写，无论项目成员写出什么样的代码，都在项目管理者的可控范围内。即使在项目中途换人，也不会有太大问题。在个人能力参差不齐的团队开发中，采用 MVC 开发是非常理想的。

（2）可以为一个模型在运行时同时建立和使用多个视图。变化-传播机制可以确保所有相关的视图及时得到模型数据变化，从而使所有关联的视图和控制器做到行为同步。

（3）视图与控制器的可接插性，允许更换视图和控制器对象，而且可以根据需求动态地打开或关闭，甚至在运行期间进行对象替换。

（4）模型的可移植性。因为模型是独立于视图的，所以可以把一个模型独立地移植到新的平台工作。需要做的只是在新平台上对视图和控制器进行新的修改。

（5）潜在的框架结构。可以基于此模型建立应用程序框架，不仅仅是用在设计界面的设计中。

MVC 模式既有优点也有缺点，总的来说还是优点大于缺点，不然也不会出现 MVC 模式。MVC 的不足体现在以下几个方面：

（1）增加了系统结构和实现的复杂性。对于简单的界面，严格遵循 MVC，使模型、视图与控制器分离，会增加结构的复杂性，并可能产生过多的更新操作，降低运行效率。

（2）视图对模型数据的访问效率低。视图可能需要多次调用 Model 才能获得足够的显示数据。

（3）完全理解 MVC 并不是很容易。使用 MVC 需要精心的计划，由于它的内部原理比较复杂，因此需要花费一些时间去思考。同时模型和视图要严格分离，给调试应用程序带来了一定的困难。

7.2 Spring MVC 处理流程

使用 Java 做 Web 开发时有好多 MVC 的框架可以使用，比如 Strut2 框架。由于 Strut2 框架被爆出有安全漏洞，现在主流的是 Spring MVC。

7.2.1 Spring MVC 引入

要学习 Spring MVC，肯定需要在项目中引入 Spring MVC。这里基于 Maven 创建一个 webapp 的工程 SpringMVCDemo，并通过 Spring MVC 在页面中输入“Hello World”，如图 7-2 所示。

默认创建的工程只有 src/main/resource 包，其他是不存在的，选中项目，右击，选择“属性”，在 build path 中可以看到有几个文件不存在，需要的话可以按照缺失的增加。如果报图 7-3 所示的错误，就需要配置 Tomcat 了。

配置 Tomcat，首先需要选中项目，然后右击，选择 properties，再单击 Java Build Path 页面中 Libraries 选项卡下的 Add Library 按钮，如图 7-4 所示。

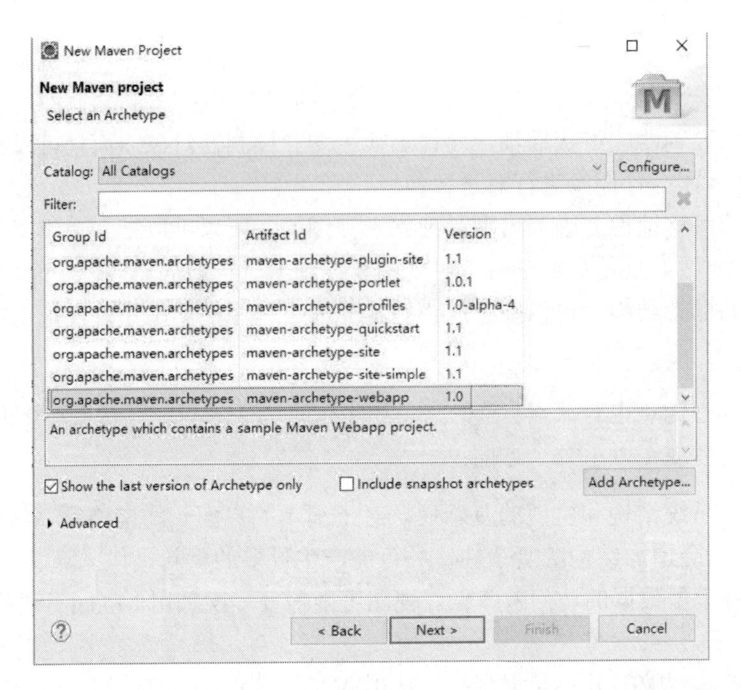

图 7-2

图 7-3

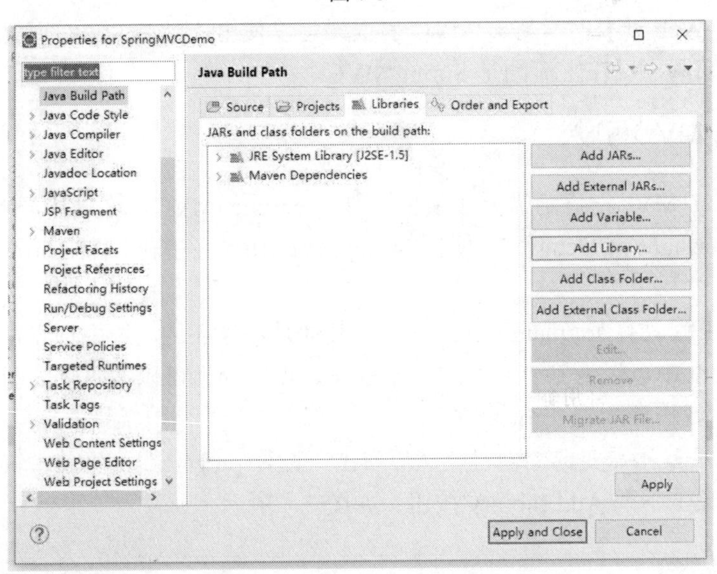

图 7-4

选择添加 Server Runtime，单击 Next 按钮，如图 7-5 所示。

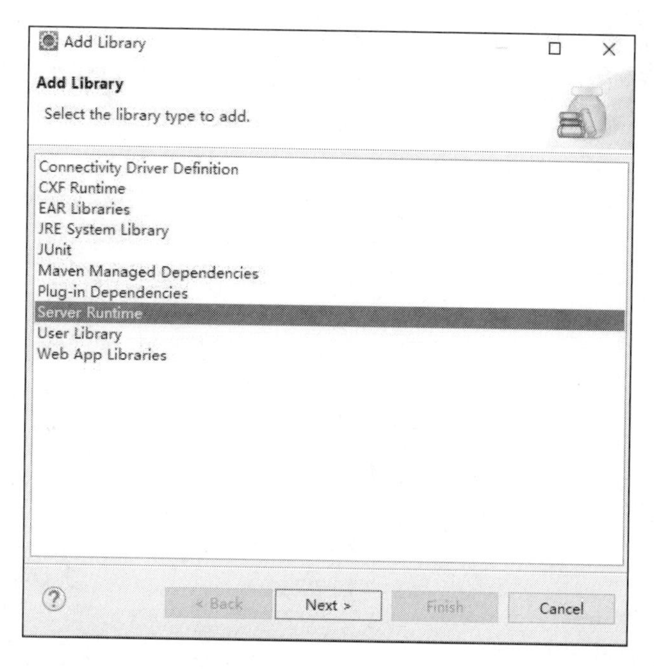

图 7-5

然后选择 Tomcat，这里选择的是 Tomcat 8.5 版本，如图 7-6 所示。

图 7-6

完整的目录结构如图 7-7 所示。

通过上面的步骤已经把项目的目录结构配置完成。下面对 Spring MVC 进行一些配置。如果对下面几个步骤中的内容不理解也没关系，这里只是先通过例子对 Spring MVC 有个整体的认识，后续会详细介绍。

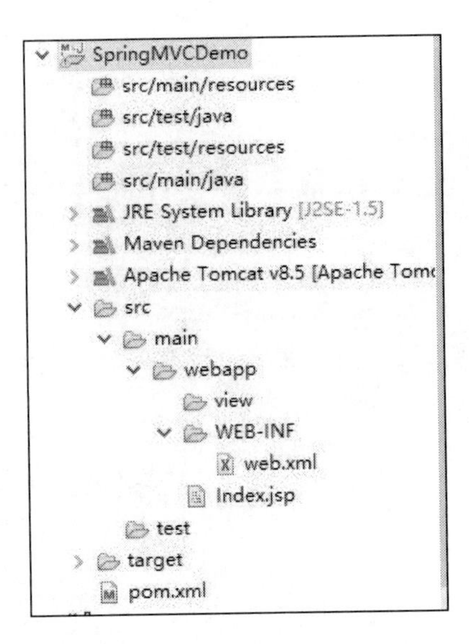

图 7-7

1. 引入依赖

```
<project xmlns="http://maven.apache.org/POM/4.0.0"
xmlns:xsi="http://www.w3.org/2001/XMLSchema-instance"
    xsi:schemaLocation="http://maven.apache.org/POM/4.0.0
http://maven.apache.org/maven-v4_0_0.xsd">
    <modelVersion>4.0.0</modelVersion>
    <groupId>com.demo</groupId>
    <artifactId>SpringMVCDemo</artifactId>
    <packaging>war</packaging>
    <version>0.0.1-SNAPSHOT</version>
    <name>SpringMVCDemo Maven Webapp</name>
    <url>http://maven.apache.org</url>
      <!-- 定义maven变量 -->
      <properties>
         <!-- spring -->
         <spring.version>5.0.0.RELEASE</spring.version>
         <!-- log -->
         <commons-logging.version>1.1.3</commons-logging.version>
         <!-- Servlet -->
         <servlet.version>3.0.1</servlet.version>
         <jsp-api.version>2.2</jsp-api.version>
         <!-- test -->
         <junit.version>3.8.1</junit.version>
         <!-- jdk -->
         <jdk.version>1.8</jdk.version>
         <jstl.version>1.2</jstl.version>
         <standard.version>1.1.2</standard.version>
         <maven.compiler.plugin.version>2.3.2
</maven.compiler.plugin.version>
```

```
</properties>
<dependencies>
    <dependency>
        <groupId>org.springframework</groupId>
        <artifactId>spring-core</artifactId>
        <version>${spring.version}</version>
    </dependency>
    <dependency>
        <groupId>org.springframework</groupId>
        <artifactId>spring-beans</artifactId>
        <version>${spring.version}</version>
    </dependency>
    <dependency>
        <groupId>org.springframework</groupId>
        <artifactId>spring-context</artifactId>
        <version>${spring.version}</version>
    </dependency>
    <dependency>
        <groupId>org.springframework</groupId>
        <artifactId>spring-jdbc</artifactId>
        <version>${spring.version}</version>
    </dependency>
    <dependency>
        <groupId>org.springframework</groupId>
        <artifactId>spring-expression</artifactId>
        <version>${spring.version}</version>
    </dependency>
    <dependency>
        <groupId>org.springframework</groupId>
        <artifactId>spring-web</artifactId>
        <version>${spring.version}</version>
    </dependency>
    <dependency>
        <groupId>org.springframework</groupId>
        <artifactId>spring-webmvc</artifactId>
        <version>${spring.version}</version>
    </dependency>
    <dependency>
        <groupId>org.springframework</groupId>
        <artifactId>spring-tx</artifactId>
        <version>${spring.version}</version>
    </dependency>
    <dependency>
        <groupId>javax.servlet</groupId>
        <artifactId>jstl</artifactId>
        <version>${jstl.version}</version>
    </dependency>
```

```
        <dependency>
            <groupId>taglibs</groupId>
            <artifactId>standard</artifactId>
            <version>${standard.version}</version>
        </dependency>
        <!-- Servlet -->
        <dependency>
            <groupId>javax.servlet</groupId>
            <artifactId>javax.servlet-api</artifactId>
            <version>${servlet.version}</version>
            <scope>provided</scope>
        </dependency>
        <dependency>
            <groupId>javax.servlet.jsp</groupId>
            <artifactId>jsp-api</artifactId>
            <version>${jsp-api.version}</version>
            <scope>provided</scope>
        </dependency>
        <!-- test -->
        <dependency>
            <groupId>junit</groupId>
            <artifactId>junit</artifactId>
            <version>${junit.version}</version>
            <scope>test</scope>
        </dependency>
    </dependencies>
    <build>
        <plugins>
            <!-- define the project compile level -->
            <plugin>
                <groupId>org.apache.maven.plugins</groupId>
                <artifactId>maven-compiler-plugin</artifactId>
                <version>${maven.compiler.plugin.version}</version>
                <configuration>
                    <source>${jdk.version }</source>
                    <target>${jdk.version}</target>
                </configuration>
            </plugin>
        </plugins>
        <finalName>SpringMVCDemo</finalName>
    </build>
</project>
```

2. 创建 Controller

在 com.demo.Controller 包里创建 IndexController。

```java
package com.demo.Controller;
import javax.servlet.http.HttpServletRequest;
import org.springframework.stereotype.Controller;
import org.springframework.web.bind.annotation.RequestMapping;
import org.springframework.web.servlet.ModelAndView;

@Controller
@RequestMapping("/index")
public class IndexController {

    @RequestMapping(value="/hello.do")

    public ModelAndView getTest(HttpServletRequest request){
        ModelAndView modelAndView = new ModelAndView("HelloWorld");
        return modelAndView;
    }
}
```

3. 创建View

在webapp/view目录下创建HelloWorld.jsp。在body中输出“Hello World!”。

```html
<%@ page language="java" contentType="text/html; charset=ISO-8859-1"
    pageEncoding="ISO-8859-1"%>
<!DOCTYPE html PUBLIC "-//W3C//DTD HTML 4.01 Transitional//EN"
"http://www.w3.org/TR/html4/loose.dtd">
<html>
<head>
<meta http-equiv="Content-Type" content="text/html; charset=ISO-8859-1">
<title>Insert title here</title>
</head>
<body>
Hello World!
</body>
</html>
```

4. 配置spring-context.xml、spring-mvc.xml

在src/main/resources包下创建spring-context.xml、spring-mvc.xml。

spring-context.xml：

```xml
<?xml version="1.0" encoding="UTF-8"?>
<beans xmlns="http://www.springframework.org/schema/beans"
xmlns:xsi="http://www.w3.org/2001/XMLSchema-instance"
xmlns:mvc="http://www.springframework.org/schema/mvc"
    xmlns:context="http://www.springframework.org/schema/context"
xmlns:tx="http://www.springframework.org/schema/tx"
xmlns:util="http://www.springframework.org/schema/util"
```

```
        xmlns:aop="http://www.springframework.org/schema/aop"
        xsi:schemaLocation="http://www.springframework.org/schema/beans
http://www.springframework.org/schema/beans/spring-beans.xsd
            http://www.springframework.org/schema/mvc
http://www.springframework.org/schema/mvc/spring-mvc.xsd
            http://www.springframework.org/schema/context
http://www.springframework.org/schema/context/spring-context.xsd
            http://www.springframework.org/schema/util
http://www.springframework.org/schema/util/spring-util.xsd
            http://www.springframework.org/schema/tx
http://www.springframework.org/schema/tx/spring-tx.xsd
            http://www.springframework.org/schema/aop
http://www.springframework.org/schema/aop/spring-aop.xsd">
        <!-- 注解注册 -->
        <context:annotation-config />
        <context:component-scan base-package="com.demo.Controller" />
    </beans>
```

spring-mvc.xml：

```
    <?xml version="1.0" encoding="UTF-8" standalone="no"?>
    <beans xmlns="http://www.springframework.org/schema/beans"
        xmlns:context="http://www.springframework.org/schema/context"
        xmlns:aop="http://www.springframework.org/schema/aop"
xmlns:mvc="http://www.springframework.org/schema/mvc"
        xmlns:p="http://www.springframework.org/schema/p"
xmlns:xsi="http://www.w3.org/2001/XMLSchema-instance"
        xsi:schemaLocation="http://www.springframework.org/schema/beans
http://www.springframework.org/schema/beans/spring-beans.xsd
            http://www.springframework.org/schema/context
http://www.springframework.org/schema/context/spring-context.xsd
            http://www.springframework.org/schema/mvc
http://www.springframework.org/schema/mvc/spring-mvc.xsd
            http://www.springframework.org/schema/aop
http://www.springframework.org/schema/aop/spring-aop.xsd">
        <!-- 自动扫描的包名 -->
        <context:component-scan base-package="com.demo.Controller" />
        <!-- 默认的注解映射的支持 -->
        <mvc:annotation-driven>
           <mvc:message-converters>
               <bean class="org.springframework.http.converter.
StringHttpMessageConverter" />
               <bean
                   class="org.springframework.http.converter.
ResourceHttpMessageConverter" />
            </mvc:message-converters>
        </mvc:annotation-driven>
```

```
    <bean id="viewResolver"
        class="org.springframework.web.servlet.view.
InternalResourceViewResolver">
        <property name="viewClass"
            value="org.springframework.web.servlet.view.JstlView" />
        <property name="prefix" value="/view/" />
        <property name="suffix" value=".jsp" />
    </bean>
    <mvc:default-servlet-handler />
</beans>
```

5. 配置 web.xml

```
<?xml version="1.0" encoding="UTF-8"?>
<web-app xmlns:xsi="http://www.w3.org/2001/XMLSchema-instance"
    xmlns="http://java.sun.com/xml/ns/javaee"
    xsi:schemaLocation="http://java.sun.com/xml/ns/javaee
http://java.sun.com/xml/ns/javaee/web-app_3_0.xsd"
    id="WebApp_ID" version="3.0">
    <display-name>SpringMVCDemo</display-name>

    <!-- 指定 Spring Bean 配置文件所在的目录，默认配置在 WEB-INF 目录下 -->
    <context-param>
        <param-name>contextConfigLocation</param-name>
        <param-value>classpath:spring-content.xml</param-value>
    </context-param>

    <!-- Spring 配置 -->
    <listener>
        <listener-class>org.springframework.web.context.
ContextLoaderListener</listener-class>
    </listener>

    <!-- Spring MVC 配置 -->
    <servlet>
        <servlet-name>spring</servlet-name>
        <servlet-class>org.springframework.web.servlet.DispatcherServlet
</servlet-class>
        <!-- 可以自定义 servlet.xml 配置文件的位置和名称，默认为 WEB-INF 目录下，名称
为[<servlet-name>]-servlet.xml，如 spring-servlet.xml -->
        <init-param>
            <param-name>contextConfigLocation</param-name>
            <param-value>classpath:spring-mvc.xml</param-value>
        </init-param>
        <load-on-startup>1</load-on-startup>
    </servlet>

    <servlet-mapping>
        <servlet-name>spring</servlet-name>
        <url-pattern>*.do</url-pattern>
```

```
        </servlet-mapping>
        <!-- 中文过滤器 -->
        <filter>
            <filter-name>CharacterEncodingFilter</filter-name>
            <filter-class>org.springframework.web.filter.
CharacterEncodingFilter</filter-class>
            <init-param>
                <param-name>encoding</param-name>
                <param-value>UTF-8</param-value>
            </init-param>
            <init-param>
                <param-name>forceEncoding</param-name>
                <param-value>true</param-value>
            </init-param>
        </filter>
        <filter-mapping>
            <filter-name>CharacterEncodingFilter</filter-name>
            <url-pattern>/*</url-pattern>
        </filter-mapping>
        <welcome-file-list>
            <welcome-file>index.jsp</welcome-file>
        </welcome-file-list>
    </web-app>
```

6. 启动运行

选中项目 run as，部署到 tomcat 中，在浏览器输入“http://localhost:8080/SpringMVCDemo/index/hello.do”，会在页面上出现“Hello World!”，如图 7-8 所示。

图 7-8

7.2.2 处理流程

不管是什么 Web 框架都要解决一个问题，就是怎么将浏览器输入的 url 字符串转换成眼前看到的 html 页面，知道它的处理流程。学习 Spring MVC 框架也是一样的，在上节的例子中通过一系列的配置在浏览器输出了“Hello World!”，为什么经过上面的配置之后在浏览器输入 url 就会有显示呢？本节了解一下 Spring MVC 的处理流程，以解决这个疑惑。一图顶千言，下面就用数据、用图说话。图 7-9 所示是 Spring MVC 的工作原理。我们可以通过下面的 11 个步骤来了解 Spring MVC 的整个流程。

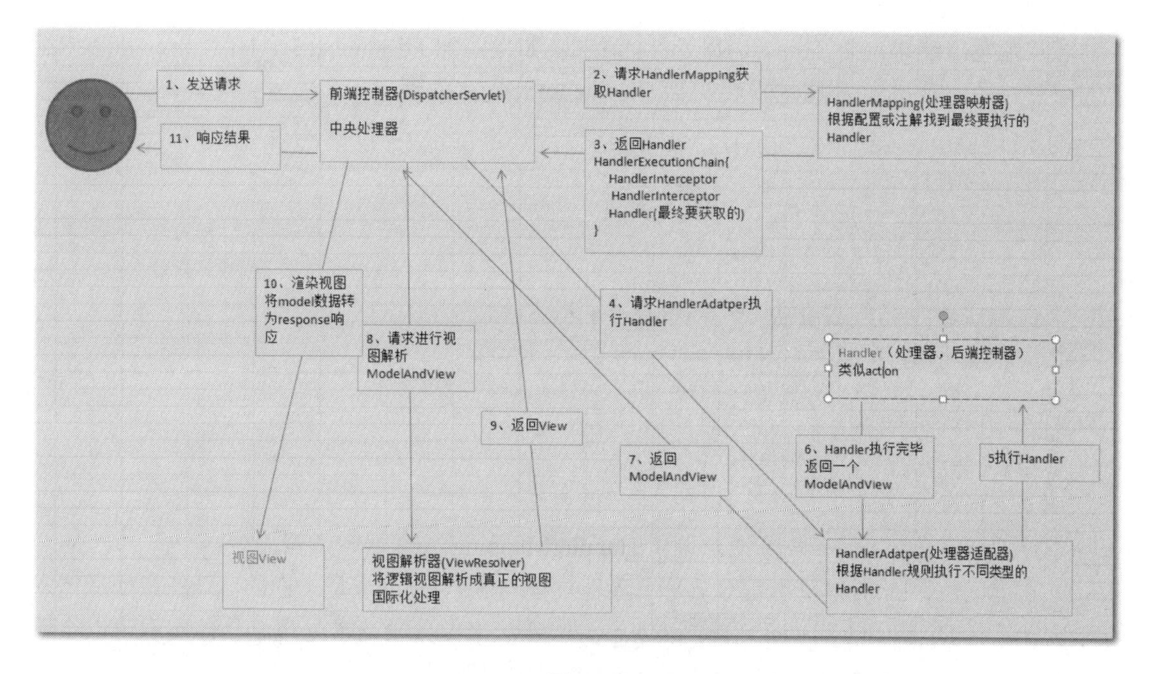

图 7-9

（1）用户发送请求至前端控制器 DispatcherServlet。

从 DispatcherServlet 的后缀也能发现 DispatcherServlet 其实就是一个 Servlet，所以在 web.xml 中的配置也与 Servlet 的配置类似。其中，<context-param>元素配置了 Servlet 容器的上下文，参数 contextConfigLocation=spring-content.xml。<servlet>元素配置了 servlet-class，值为 org.springframework.web.servlet.DispatcherServlet，初始化参数 contextConfigLocation 指向 spring-mvc.xml，作为 Servlet 的上下文。这里 spring-content.xml 是 Servlet 容器的上下文，spring-mvc.xml 是 Servlet 的上下文（也是 Spring MVC 框架的上下文）。在 spring-mvc.xml 中可以配置自动扫描包名、默认注解映射支持、视图解释类、拦截器、对静态资源文件的访问等信息，这些会在 Tomcat 进行参数初始化的时候实例化完成。在 DispatcherServlet 中维护着一个表，存放的是 HandlerMapping 对象 list。下面截取的是 DispatcherServlet 中的部分代码，在 DispatcherServlet 中维护着 handerMappings、handerAdapters 等对象列表。在 initStrategies 中对上面的一些属性进行初始化。

```
15      private LocaleResolver localeResolver;
16
17      /** ThemeResolver used by this servlet */
18      @Nullable
19      private ThemeResolver themeResolver;
20
21      /** List of HandlerMappings used by this servlet */
22      @Nullable
23      private List<HandlerMapping> handlerMappings;
24
25      /** List of HandlerAdapters used by this servlet */
26      @Nullable
27      private List<HandlerAdapter> handlerAdapters;
28
29      /** List of HandlerExceptionResolvers used by this servlet */
30      @Nullable
31      private List<HandlerExceptionResolver> handlerExceptionResolvers;
32
33      /** RequestToViewNameTranslator used by this servlet */
34      @Nullable
35      private RequestToViewNameTranslator viewNameTranslator;
```

客户端发出请求，由 Tomcat 接收这个请求，如果匹配 DispatcherServlet 在 web.xml 中配置的映射路径，Tomcat 就将请求转交给 DispatcherServlet 处理。

```
protected void initStrategies(ApplicationContext context) {
    initMultipartResolver(context);
    initLocaleResolver(context);
    initThemeResolver(context);
    initHandlerMappings(context);
    initHandlerAdapters(context);
    initHandlerExceptionResolvers(context);
    initRequestToViewNameTranslator(context);
    initViewResolvers(context);
    initFlashMapManager(context);
}
```

（2）DispatcherServlet 收到请求调用 HandlerMapping 处理器映射器。

DispatcherServlet 的本质也是 Servlet，运行过程与 Servlet 请求是一样的，只是在处理请求的时候会调用 DispatcherServlet 中的 doDispatch 方法来做分发请求处理。具体实现可以查看 org.springframework.web.servlet.DispatcherServlet 里面的源码。

```
    protected void doDispatch(HttpServletRequest request,
HttpServletResponse response) throws Exception {
        HttpServletRequest processedRequest = request;
        HandlerExecutionChain mappedHandler = null;
        boolean multipartRequestParsed = false;

        WebAsyncManager asyncManager = WebAsyncUtils.getAsyncManager
(request);

        try {
            ModelAndView mv = null;
            Exception dispatchException = null;

            try {
                processedRequest = checkMultipart(request);
                multipartRequestParsed = (processedRequest != request);

                // Determine handler for the current request.
                mappedHandler = getHandler(processedRequest);
                if (mappedHandler == null) {
                    noHandlerFound(processedRequest, response);
                    return;
                }

                // Determine handler adapter for the current request.
                HandlerAdapter ha = getHandlerAdapter(mappedHandler.
getHandler());
```

```
                    // Process last-modified header, if supported by the handler.
                    String method = request.getMethod();
                    boolean isGet = "GET".equals(method);
                    if (isGet || "HEAD".equals(method)) {
                        long lastModified = ha.getLastModified(request,
mappedHandler.getHandler());
                        if (logger.isDebugEnabled()) {
                            logger.debug("Last-Modified value for [" +
getRequestUri(request) + "] is: " + lastModified);
                        }
                        if (new ServletWebRequest(request,
response).checkNotModified(lastModified) && isGet) {
                            return;
                        }
                    }

                    if (!mappedHandler.applyPreHandle(processedRequest,
response)) {
                        return;
                    }

                    // Actually invoke the handler.
                    mv = ha.handle(processedRequest, response,
mappedHandler.getHandler());

                    if (asyncManager.isConcurrentHandlingStarted()) {
                        return;
                    }

                    applyDefaultViewName(processedRequest, mv);
                    mappedHandler.applyPostHandle(processedRequest, response,
mv);
                }
                catch (Exception ex) {
                    dispatchException = ex;
                }
                catch (Throwable err) {
                    // As of 4.3, we're processing Errors thrown from handler
methods as well,
                    // making them available for @ExceptionHandler methods and
other scenarios.
                    dispatchException = new NestedServletException("Handler
dispatch failed", err);
                }
               processDispatchResult(processedRequest, response, mappedHandler,
mv, dispatchException);
            }
            catch (Exception ex) {
```

```
                triggerAfterCompletion(processedRequest, response,
mappedHandler, ex);
            }
            catch (Throwable err) {
                triggerAfterCompletion(processedRequest, response,
mappedHandler,
                new NestedServletException("Handler processing failed", err));
            }
            finally {
                if (asyncManager.isConcurrentHandlingStarted()) {
                    // Instead of postHandle and afterCompletion
                    if (mappedHandler != null) {
                        mappedHandler.applyAfterConcurrentHandlingStarted
(processedRequest, response);
                    }
                }
                else {
                    // Clean up any resources used by a multipart request.
                    if (multipartRequestParsed) {
                        cleanupMultipart(processedRequest);
                    }
                }
            }
        }
```

（3）处理器映射器找到具体的处理器 Handler。

处理器映射器找到具体的处理器 Handler（可以根据 xml 配置、注解进行查找，这里的 Handler 其实指的就是 Controller），生成处理器对象及处理器拦截器（如果有则生成）一并返回给 DispatcherServlet。

```
            mappedHandler = getHandler(processedRequest);
            if (mappedHandler == null) {
                noHandlerFound(processedRequest, response);
                return;
            }
```

上面的代码调用 getHandler 方法来获取具体的处理器，下面给出 getHandler 的具体实现，从中可以看出它返回了一个 HandlerExecutionChain 类型的对象。

```
179      */
180     @Nullable
181     protected HandlerExecutionChain getHandler(HttpServletRequest request) throws Exception {
182         if (this.handlerMappings != null) {
183             for (HandlerMapping hm : this.handlerMappings) {
184                 if (logger.isTraceEnabled()) {
185                     logger.trace(
186                             "Testing handler map [" + hm + "] in DispatcherServlet with name '" + getServletName() + "'");
187                 }
188                 HandlerExecutionChain handler = hm.getHandler(request);
189                 if (handler != null) {
190                     return handler;
191                 }
192             }
193         }
194         return null;
195     }
```

HandlerExecutionChain 对象包含一个 handler 和多个 HandlerInterceptor（拦截器）。这和 ASP.NET 中的管道模型有点类似，一个 httphandler 和多个 httpmodule。httpmodule 也是用来做拦截操作的，我们可以转到定义中查看 HandlerExecutionChain 的数据结构（如下所示）。Spring MVC 的拦截器实现原理就是通过 HandlerExecutionChain 对象的 HandlerInterceptor 进行拦截。

```
37   * @see HandlerInterceptor
38   */
39  public class HandlerExecutionChain {
40
41      private static final Log logger = LogFactory.getLog(HandlerExecutionChain.class);
42
43      private final Object handler;
44
45      @Nullable
46      private HandlerInterceptor[] interceptors;
47
48      @Nullable
49      private List<HandlerInterceptor> interceptorList;
50
51      private int interceptorIndex = -1;
52
53
54      /**
55       * Create a new HandlerExecutionChain.
56       * @param handler the handler object to execute
57       */
58      public HandlerExecutionChain(Object handler) {
59          this(handler, (HandlerInterceptor[]) null);
```

（4）DispatcherServlet 调用 HandlerAdapter 处理器适配器。

前面几步通过映射找到对应的 Controller 和拦截器，是找的过程，找到之后需要先将处理器 Handler 包装成适配器，这样就可以支持多种类型的处理器，然后就是执行的过程。

```
HandlerAdapter ha = getHandlerAdapter(mappedHandler.getHandler());
```

（5）HandlerAdapter 经过适配调用具体的处理器 Handler（Controller，也叫后端控制器）。

（6）Controller 执行完成，返回 ModelAndView。

（7）HandlerAdapter 将 controller 执行结果 ModelAndView 返回给 DispatcherServlet。

```
String method = request.getMethod();
boolean isGet = "GET".equals(method);
if (isGet || "HEAD".equals(method)) {
   long lastModified = ha.getLastModified(request, mappedHandler.getHandler());
   if (logger.isDebugEnabled()) {
       logger.debug("Last-Modified value for [" + getRequestUri(request) +
"] is: " + lastModified);
   }
   if (new ServletWebRequest(request, response).checkNotModified
(lastModified) && isGet) {
       return;
   }
}

if (!mappedHandler.applyPreHandle(processedRequest, response)) {
   return;
}
```

```
    // Actually invoke the handler.
    mv = ha.handle(processedRequest, response, mappedHandler.getHandler());
```

上面的几行代码先判断如果是 get 请求就更新 lastModified 请求头，然后执行 HandlerExecutionChain 中的 applyPreHandle 这个方法。我们来看一下 applyPreHandle 的具体实现以及 applyPreHandle 方法中调用的 triggerAfterCompletion 方法。

```
        boolean applyPreHandle(HttpServletRequest request, HttpServletResponse
 response) throws Exception {
            HandlerInterceptor[] interceptors = getInterceptors();
            if (!ObjectUtils.isEmpty(interceptors)) {
                for (int i = 0; i < interceptors.length; i++) {
                    HandlerInterceptor interceptor = interceptors[i];
                    if !interceptor.preHandle(request, response, this.handler)){
                        triggerAfterCompletion(request, response, null);
                        return false;
                    }
                    this.interceptorIndex = i;
                }
            }
            return true;
        }
        void triggerAfterCompletion(HttpServletRequest request,
 HttpServletResponse response, @Nullable Exception ex)
                throws Exception {

            HandlerInterceptor[] interceptors = getInterceptors();
            if (!ObjectUtils.isEmpty(interceptors)) {
                for (int i = this.interceptorIndex; i >= 0; i--) {
                    HandlerInterceptor interceptor = interceptors[i];
                    try {
                        interceptor.afterCompletion(request, response,
 this.handler, ex);
                    }
                    catch (Throwable ex2) {
                        logger.error("HandlerInterceptor.afterCompletion threw
 exception", ex2);
                    }
                }
            }
        }
```

在 applyPreHandle 中它会遍历该 HandlerExecutionChain 中所有的拦截器，然后使用拦截器通过 preHandle 方法对 handler 进行预处理，如果所有的拦截器都能处理就继续往下执行，一旦有一个拦截器不能处理，就没必要往下走了，就会触发 triggerAfterCompletion 方法。在 triggerAfterCompletion 中它是倒序遍历拦截器的，执行完 triggerAfterCompletion 返回 false 之后，doDispatch 方法就执行结束了，8~11 步就不再执行了。

（8）DispatcherServlet 将 ModelAndView 传给 ViewReslover 视图解析器。

（9）ViewReslover 解析后返回具体 View。

```
applyDefaultViewName(processedRequest, mv);
mappedHandler.applyPostHandle(processedRequest, response, mv);
```

在 applyDefaultViewName 中会找到 ModelAndView 对应的 viewname，设置成它的属性值。在 applyPostHandle 中会执行拦截器的 postHandle 方法。这里的 applyPostHandle 与第 7 步的 applyPreHandle 分别会执行 HandlerInterceptor 的 postHandle 和 preHandle 方法。自定义拦截器里面的方法就是在第 7 步和第 9 步被调用的。

（10）DispatcherServlet 根据 View 进行渲染视图（将模型数据填充至视图中）。

```
    processDispatchResult(processedRequest, response, mappedHandler, mv,
dispatchException);
```

在 processDispatchResult 方法中有一句 render(mv, request, response)，在 render 方法中先是获取 View 对象，然后调用“view.render(mv.getModelInternal(), request, response);”，将 view 和 model 绑定来渲染视图。

（11）DispatcherServlet 响应用户。

在父类 FrameworkServlet 的 processRequest 方法中执行 publishRequestHandledEvent(request, response, startTime, failureCause)方法，将结果响应给用户。

7.3 HandlerMapping 的使用

在 com.demo.Controller 包的 IndexController 类中使用注解@RequestMapping 配置在 IndexController 类和 getTest 方法上之后就可以在 url 中输入@RequestMapping 注解的值执行 getTest 方法，那它是怎么实现的呢？这就要了解 HandlerMapping 的使用了。图 7-10 所示是 HandlerMapping 接口的 API 文档，可以看到该接口有好几个实现类。

org.springframework.web.servlet

Interface HandlerMapping

All Known Subinterfaces:

MatchableHandlerMapping

All Known Implementing Classes:

AbstractDetectingUrlHandlerMapping, AbstractHandlerMapping, AbstractHandlerMethodMapping, AbstractUrlHandlerMapping, BeanNameUrlHandlerMapping, RequestMappingHandlerMapping, RequestMappingInfoHandlerMapping, SimpleUrlHandlerMapping, WebSocketHandlerMapping

图 7-10

HandlerMapping 接口几个实现类之间的关系可以用 UML 类图表示出来，如图 7-11 所示。

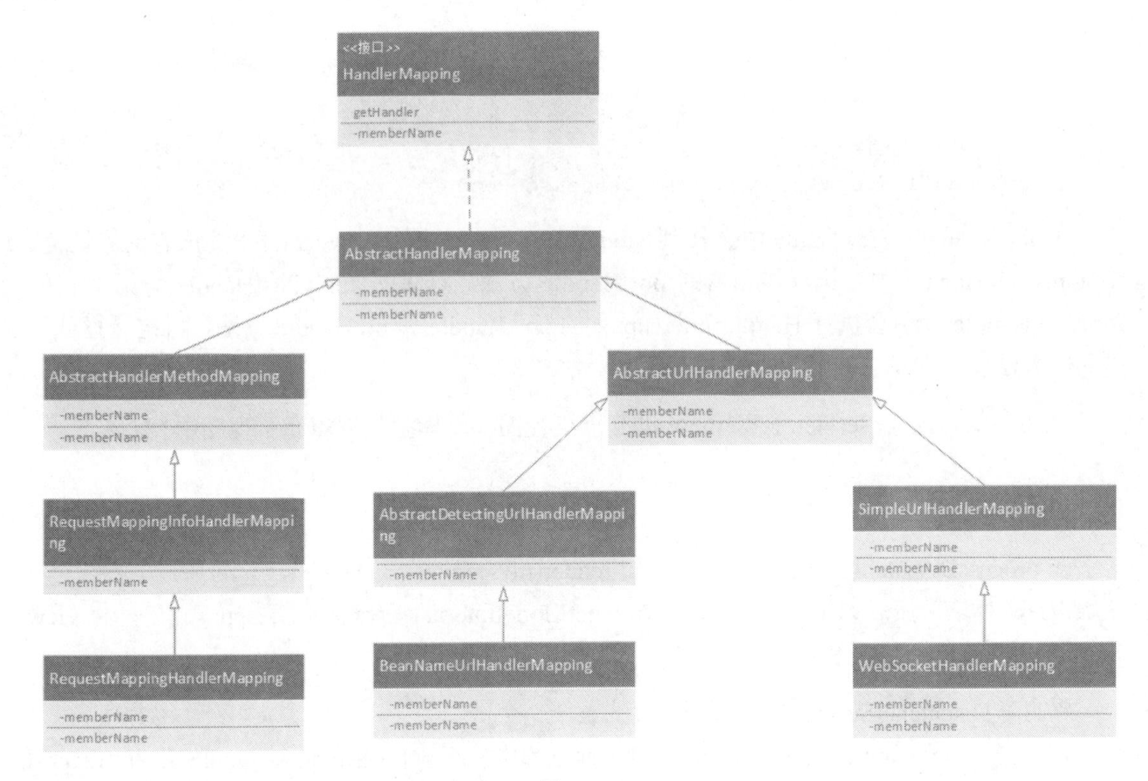

图 7-11

在 HandlerMapping 的实现类中常用的有 3 个，分别是 RequestMappingHandlerMapping、BeanNameUrlHandlerMapping 和 SimpleUrlHandlerMapping。

7.3.1 RequestMappingHandlerMapping

RequestMappingInfoHandlerMapping 和 AbstractHandlerMethodMapping 都是抽象类，实际上最终还是用的 RequestMappingHandlerMapping，使用注解的时候，通过注解将 url 映射到对应的 Controller 上。

```
/**
 * Creates {@link RequestMappingInfo} instances from type and method-level
 * {@link RequestMapping @RequestMapping} annotations in
 * {@link Controller @Controller} classes.
 *
 * @author Arjen Poutsma
 * @author Rossen Stoyanchev
 * @author Sam Brannen
 * @since 3.1
 */
public class RequestMappingHandlerMapping extends
RequestMappingInfoHandlerMapping
        implements MatchableHandlerMapping, EmbeddedValueResolverAware {
```

在上面的 IndexController 类中，就是使用的 RequestMappingHandlerMapping 来将 url 映射到 getTest 方法上的。

7.3.2 BeanNameUrlHandlerMapping

BeanNameUrlHandlerMapping 通过 bean 将 url 映射到对应的 Controller，需要在 spring-mvc.xml 中配置 BeanNameUrlHandlerMapping 以及 bean 对应的 Controller。Controller 要继承 AbstractController。

```
package com.demo.Controller;

import javax.servlet.http.HttpServletRequest;
import javax.servlet.http.HttpServletResponse;

import org.springframework.web.servlet.ModelAndView;
import org.springframework.web.servlet.mvc.AbstractController;

public class HelloController extends AbstractController{

    @Override
    protected ModelAndView handleRequestInternal(HttpServletRequest request,
HttpServletResponse response)
            throws Exception {

        return new ModelAndView("HelloWorld");
    }
}
```

在 spring-mvc.xml 中配置 bean，将 url 映射到对应的 Controller。

```
    <bean class="org.springframework.web.servlet.handler.
BeanNameUrlHandlerMapping"></bean>
    <bean id="/hello.do" class="com.demo.Controller.HelloController">
</bean>
```

在浏览器中输入“url：http://localhost:8080/SpringMVCDemo/hello.do”，可以将 HelloWorld.jsp 页面的内容输出，如图 7-12 所示。

图 7-12

7.3.3 SimpleUrlHandlerMapping

BeanNameUrlHandlerMapping 是通过 bean 的名字来实现 handler 映射的。SimpleUrlHandlerMapping 则是通过 url 来实现 handler 的映射，有两种常见的映射方法：一是通过 prop key，二是通过 value。

这里还是使用 HelloController，在 spring-mvc.xml 中的配置可以使用 props 元素来进行配置。prop 标签中的 key 都是 url 值，后面的为 BeanID，如果我们在地址栏中输入的 url 与 key 匹配，则分发到 prop 标签中指定的 beanID 所指定的 Controller。

```
<bean id="helloController" class="com.demo.Controller.HelloController" />
<bean class="org.springframework.web.servlet.handler.
SimpleUrlHandlerMapping">
    <property name="mappings">
     <props>
        <prop key="/hello.html">helloController</prop>
      </props>
    </property>
</bean>
```

value 元素和 prop key 类似，通过 url=bean 的形式将浏览器输入的 url 与 xml 中配置的 Controller 进行映射。

```
<bean id="helloController" class="com.demo.Controller.HelloController" />
<bean class="org.springframework.web.servlet.handler.
SimpleUrlHandlerMapping">
        <property name="mappings">
         <value>
            /*/hello.html=helloController
         </value>
        </property>
</bean>
```

两种方式都可以通过输入"http://localhost:8080/SpringMVCDemo/hello.html"将 HelloWorld.jsp 页面的内容输出。

7.4 传递数据到 Controller

MVC 框架都要解决 url 映射问题、数据交互问题，上一节介绍了 url 映射问题，从本节开始来介绍数据交互问题。MVC 中数据传递主要分为从 url 到 Controller、从 view 到 Controller、Controller 到 view 以及 Controller 之间的数据传递几个部分。

7.4.1 URL 传递数据到 Controller

在浏览器输入的url有时会带问号，问号之后用&取地址符连接name=value，在Spring MVC 中这些参数 Controller 是用什么接收的呢？Controller 接收参数主要用@RequestParam 和@PathVariable。这里在 webapp/view 目录下新建 index.jsp 页面，在里面添加一个表单，用来输入用户名和密码。

```
<%@ page language="java" contentType="text/html; charset=utf-8"
    pageEncoding="utf-8"%>
<!DOCTYPE html PUBLIC "-//W3C//DTD HTML 4.01 Transitional//EN"
"http://www.w3.org/TR/html4/loose.dtd">
<html>
<head>
<meta http-equiv="Content-Type" content="text/html; charset=utf-8">
<title>Insert title here</title>
</head>
<body>

<form id="form1" name="myform" method="post" action="login.do" >
     用户:<input type="text" name="name"/>
     密码:<input type="password" name="pwd"/>
     <input type="submit"/>
 </form>
</body>
</html>
```

1. @RequestParam

常见的url中会是"?name=XXX&pwd=XXX"这种，如果想获取"name,pwd"，可以使用@RequestParam。假如是可选参数，可以设置required=false，默认是true，value也要与url的对应。

```
    @RequestMapping(value="/login.do")
    public ModelAndView login(HttpServletRequest request,
HttpServletResponse response,@RequestParam("name")String name,
@RequestParam(value="pwd",required=false) String pwd){
        ModelAndView modelAndView = new ModelAndView("Index");
        System.out.println("name:"+name+" pwd:"+pwd);
        return modelAndView;
        }
```

在浏览器中输入"http://localhost:8080/SpringMVCDemo/index/login.do?name=test&pwd=123456"，Controller就会将name和pwd输出到控制台，并将index.jsp页面显示出来。

2. @PathVariable

有的url的格式是url/param1/param2，这种获取值可以使用@PathVariable。

```
    @RequestMapping(value="/getlogin/{name}/{pwd}",
method=RequestMethod.GET)
    public ModelAndView getlogin(HttpServletRequest request,
HttpServletResponse response,@PathVariable("name") String name,
@PathVariable("pwd") String pwd){
        ModelAndView modelAndView = new ModelAndView("Index");
        System.out.println("name:"+name+" pwd:"+pwd);
        return modelAndView;
        }
```

在浏览器中输入“http://localhost:8080/SpringMVCDemo/index/getlogin/test/123456”，Controller 同样会将 name 和 pwd 输出到控制台并显示 index.jsp 页面。

7.4.2 View 传递数据到 Controller

上面是通过 url 传递参数到 Controller 中，通常用户是通过页面 View 填充数据之后发送到服务端，从 View 传递数据到 Controller 主要有 3 种方式：直接将请求参数名作为 Controller 中方法的形参、使用 Pojo 对象和使用原生的 Servlet API 作为 Controller 方法的参数。

1. 直接将请求参数名作为 Controller 中方法的形参

这里定义了 login2 方法，用于接收 login.do 的 post 请求，形参 name、pwd 用来接收表单提交到 Controller 中的数据。在 index.jsp 中输入用户名、密码之后，单击“提交”按钮，在控制台中可以看到从 view 中传入的 name 和 pwd。

```
    @RequestMapping(value="/login.do",method=RequestMethod.POST)
    public ModelAndView login2(HttpServletRequest request,
HttpServletResponse response,String name,String pwd){
        ModelAndView modelAndView = new ModelAndView("Index");
        System.out.println("name:"+name+" pwd:"+pwd);
        return modelAndView;
        }
```

2. 使用 Pojo 对象

如果请求参数有好多个，都放在 Controller 方法的形参中时函数会很长，一般情况下会把参数封装成对象进行传递。Spring MVC 会按请求参数名和 POJO 属性名进行自动匹配，自动为该对象填充属性值，支持级联属性，如 address.province、address.city 等。

为了演示如何使用 Pojo 对象，首先在 com.demo.model 包中定义 User 对象，该对象中有两个属性—— name 和 pwd。

```
package com.demo.model;
public class User {
    private String name;
    private String pwd;
    public String getName() {
        return name;
    }
    public void setName(String name) {
        this.name = name;
    }
    public String getPwd() {
        return pwd;
    }
    public void setPwd(String pwd) {
```

```
            this.pwd = pwd;
        }
    }
```

在 Controller 中可以使用 User 来接收 View 传过来的参数。

```
    @RequestMapping(value="/login.do",method=RequestMethod.POST)
    public ModelAndView login3(HttpServletRequest
request,HttpServletResponse response,User user){
        ModelAndView modelAndView = new ModelAndView("Index");
        System.out.println("name:"+user.getName()+" pwd:"+user.getPwd());
        return modelAndView;
        }
```

3. 使用原生的 Servlet API 作为 Controller 方法的参数

不仅是通过view传到Controller，url传参数到Controller也一样。既然有HttpServletRequest，我们可以通过 request 对象获取相关参数。

```
        String username=request.getParameter("name");
        System.out.println("username:"+username);
        String pwd=request.getParameter("pwd");
        System.out.println("pwd:"+pwd);
```

7.5 传递数据到 View

上一节介绍了如何从 url 或 view 中向 Controller 传递数据，是用户输入的过程，本节介绍如何从 Controller 向 View 传递数据，是向用户输出的过程。数据主要通过 ModelAndView、Model / Map / ModelMap、@SessionAttributes、@ModelAttribute 这几种方式传递给 View。

7.5.1 ModelAndView

在前面 Controller 方法中都是通过 new ModelAndView("Index")仅是返回视图，其实如果转到 ModelView 类里面可以看到该类还有一个继承自 LinkedHashMap 的 ModelMap 类型的 model 属性，可以通过该属性来传参数到 View 中。

```
    @RequestMapping(value="/login.do",method=RequestMethod.POST)
    public ModelAndView login4(HttpServletRequest request,
HttpServletResponse response,User user){
        ModelAndView modelAndView = new ModelAndView("Index");
        modelAndView.addObject("user", user);
        return modelAndView;
        }
```

使用 addObject 方法，往 hashmap 中 put 了一个 key=user，value=user 的对象，之后就可以在 View 中获取到。在 index.jsp 的 body 体中增加下面两行获取 user 对象属性的代码，当用户在页面提交之后就可以看到提交的内容，如图 7-13 所示。

```
姓名:${user.name}<br>
密码:${user.pwd}<br>
```

图 7-13

在 Controller 中的方法入参如果为 org.springframework.ui.Model、Java.uti.Map 或 org.springframework.ui.ModelMap，在方法返回时会将参数自动添加到模型中。

```
@RequestMapping(value = "/testModel.do",method = RequestMethod.GET)
public String testModel(Model model){
      User user=new User();
      user.setName("abc");
      user.setPwd("123456");
      model.addAttribute("user",user);
      return "Index";
}
```

在浏览器中输入“http://localhost:8080/SpringMVCDemo/index/testModel.do”，会返回 Index.jsp 页面，同时会将 user 对象传递到 View 中，如图 7-14 所示。

图 7-14

7.5.2 @SessionAttributes

@SessionAttributes 会将模型中的某个属性暂存到 HttpSession 中，以便多个请求之间可以共享这个属性，比如我们登录之后可能会存登录信息，就可以使用它。@SessionAttributes 这个注解只能放到类的上面。

```
@SessionAttributes({"user"})
@Controller
@RequestMapping("/index")
public class IndexController {
```

在 IndexController 上面配置了@SessionAttributes，将 attributeName 为 user 的对象保存在 HttpSession 中，这样其他地方也能获取到该对象。

在 HelloWorld.jsp 页面的 body 中也加上 user 对象的 name、pwd 属性的代码。

```
姓名:${user.name}<br>
密码:${user.pwd}<br>
```

在浏览器中输入“http://localhost:8080/SpringMVCDemo/index/testModel.do”，将 user 对象保存在 HttpSession 中，然后输入 url“http://localhost:8080/SpringMVCDemo/index/test.do”，虽然只是返回视图，但是显示的页面也可以获取 user 对象的 name 和 pwd，如图 7-15 所示。

```
@RequestMapping(value="/test.do")
public ModelAndView getTest(HttpServletRequest request){
    ModelAndView modelAndView = new ModelAndView("HelloWorld");
    return modelAndView;
}
```

图 7-15

7.5.3 @ModelAttribute

Spring MVC 在每次调用请求处理方法时都会创建 Model 类型的一个实例。如果准备使用此实例，就可以在方法中添加一个 Model 类型的参数。可以用@ModelAttribute 来注释方法参数：带有@ModelAttribute 注解的方法会将其输入或创建的参数对象添加到 Model 对象中（若方法中没有显式添加）。也可以用@ModelAttribute 标注一个非请求的处理方法（有返回值，无返回值）：被@ModelAttribute 注释的方法会在此 Controller 每个方法执行前被执行。

1. @ModelAttribute 注解方法参数

```
    @RequestMapping(value = "/testModelAttribute.do",method = 
RequestMethod.GET)
    public String testModelAttribute(HttpServletRequest request, 
HttpServletResponse response,@ModelAttribute("user")User 
user,@ModelAttribute("name")String name,Model model){
        user.setName("abc");
        user.setPwd("123456");
        name="testModelAttribute";
        model.addAttribute("name", name);
        return "Index";
    }
```

使用@ModelAttribute 注解 user 和 name 之后，在 testModelAttribute 方法中修改了参数的值，然后在 index.jsp 显示出来，这里在 index.jsp 中增加了 name 属性的获取，在浏览器中输入"http://localhost:8080/SpringMVCDemo/index/testModelAttribute.do"，结果如图 7-16 所示。

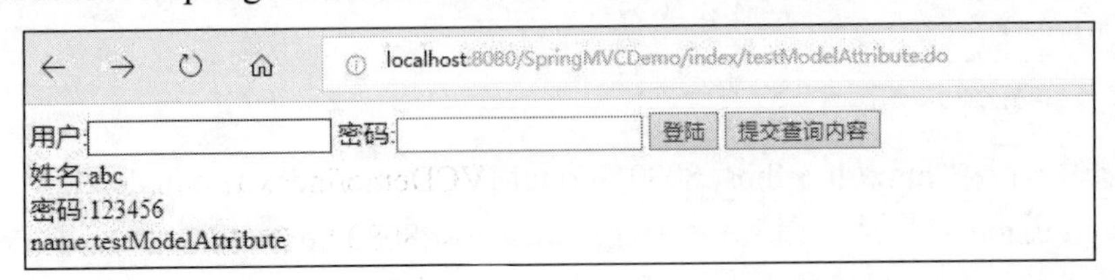

图 7-16

2. @ModelAttribute 标注一个非请求的处理方法

被@ModelAttribute 注解的方法会在此 Controller 每个方法执行前被执行。被@ModelAttribute 注解的方法还区分有返回值和无返回值两种。

（1）有返回值方法

在 IndexController 中增加了使用@ModelAttribute 注解的方法，在 map()方法里面实例化了 User 对象，并将该对象放在 HashMap 中返回。

```
@ModelAttribute(value = "mymap")
public Map<String,Object> map() {
  User user=new User();
  user.setName("abc");
  user.setPwd("123456");
  HashMap<String, Object> map=new HashMap<String, Object>();
  map.put("user",user);
  return map;
}
```

在 jsp 页面中获取就不能直接用${user.name}来获取值，还要在前面加上@ModelAttribute 的 value 值。在浏览器中输入 url"http://localhost:8080/SpringMVCDemo/index/test.do"，通过 map()方法设置的对象就能获取显示出来，如图 7-17 所示。

```
姓名:${mymap.user.name}<br>
密码:${mymap.user.pwd}<br>
```

图 7-17

（2）无返回值方法

有返回值和无返回值最大的区别是无返回值的方法需要显式地将对象添加到 model 中。

```
@ModelAttribute
public void voidmap(Model model) {
  User user=new User();
  user.setName("abcd");
  user.setPwd("123456");
  HashMap<String, Object> map=new HashMap<String, Object>();
  map.put("user",user);
  model.addAttribute("voidmap",map);
}
```

在 IndexController 中又增加了一个无返回值的方法 voidmap，在方法中显式为 model 注入 HashMap 对象。然后在 HelloWorld.jsp 中修改一下获取值，代码如下：

```
姓名:${voidmap.user.name}<br>
密码:${voidmap.user.pwd}<br>
```

登录之后，在浏览器中输入 url“http://localhost:8080/SpringMVCDemo/index/test.do”，则会将 voidmap 中的 model 对象显示在 HelloWorld.jsp 页面中，如图 7-18 所示。

图 7-18

7.6 拦截器的使用

在介绍 Spring MVC 工作流程时介绍了 HandlerExecutionChain 对象包含一个 handler 和多个 HandlerInterceptor（拦截器）。本节通过实例了解一下 Spring MVC 中拦截器的使用。

我们平时在做系统时有些页面是需要先登录才能访问的，一种方法是在每个请求方法中都做登录判断，这样顶多是把登录功能封装起来，以后每新增一个代码都要加上，很不方便。其实这里我们可以使用拦截器进行登录验证，判断是否有 session，如果有 session 就断定已经登录。拦截器不仅仅可以做登录，登录完成之后可能还有根据用户角色限制页面或工具的权限，我们还可以再增加一个拦截器来判断用户权限等。本节使用登录来演示拦截器的使用。

1. 新建拦截器类

在 com.demo.model 包下新建拦截器 LoginIntercepter 类，实现 HandlerInterceptor 接口，重写了 preHandle 方法。在 preHandle 方法中判断是不是登录页面，登录页面不能被拦截，不然它始终登录不上。然后判断是否有 session，如果有就当作登录成功，没有就跳转到登录页面。

```
package com.demo.model;
import javax.servlet.http.HttpServletRequest;
```

```
import javax.servlet.http.HttpServletResponse;
import javax.servlet.http.HttpSession;
import org.springframework.web.servlet.HandlerInterceptor;
public class LoginIntercepter implements HandlerInterceptor{

    @Override
    public boolean preHandle(HttpServletRequest request, HttpServletResponse
response, Object handler)
            throws Exception {
        String requestURI = request.getRequestURI();
        if(requestURI.indexOf("/login")<=0){
            HttpSession session = request.getSession();
            User user=  (User) session.getAttribute("user");
            if(user!=null){
                //登录成功的用户
                return true;
            }else{
                //没有登录，转向登录界面
                request.getRequestDispatcher("../view/Index.jsp").
forward(request,response);
                return false;
            }
        }else{
            return true;
        }
    }
}
```

这里在进行页面跳转时使用的是 forward 方法，还有一种页面跳转方法 forward，下面给出两者的区别。

- forward 过程：转发，服务器端行为。Web 服务器接受请求，调用内部的方法在容器内部完成请求处理和转发动作，然后响应客户端，在这里转发的路径必须是同一个 web 容器下的 url，不能转向其他的 Web 路径上去，中间传递的是自己容器内的 request。
- redirect 过程：重定向，客户端行为。客户端发送 HTTP 请求，Web 服务器接受后发送 3** 状态码响应及对应新的 location 给客户端，客户端发现是 3**响应，则自动发送一个新的 http 请求，请求 url 是新的 location 地址，在这里 location 可以重定向到任意 URL，既然是浏览器重新发出了请求，就没有什么 request 传递的概念了。重定向行为是浏览器做了至少两次的访问请求。

2. 配置拦截器

创建之后还需要将拦截器配置到项目中，需要在 spring-mvc.xml 中使用<mvc:interceptors>元素将创建的拦截器类配置到 HandlerExecutionChain 中。

```
<mvc:interceptors>
    <mvc:interceptor><mvc:mapping path="/**"/><bean
```

```
class="com.demo.model.LoginIntercepter"/></mvc:interceptor>
        </mvc:interceptors>
```

3. 测试

在 IndexController 中使用@SessionAttributes({"user"})设置了 session 对象，当第一次在浏览器中输入 url“http://localhost:8080/SpringMVCDemo/index/test.do”时会自动跳转到 Index.jsp 页面。在 Index.jsp 页面输入用户名、密码提交之后再次输入 url“http://localhost:8080/SpringMVCDemo/index/test.do”就会正常显示，不会再跳转到登录页面，如图 7-19 所示。

图 7-19

7.7 Ajax 与 Controller 交互

Ajax 与 Controller 交互主要有 3 个步骤，引入静态资源文件、Ajax 发起请求、Controller 接收请求并响应。

1. 引入静态资源文件

前面学习了拦截器，通过拦截器我们可以拦截请求，做进一步处理之后再往下进行，但是我们使用 Ajax 的时候会有一个问题，就是会把 js、css 这些静态资源文件拦截了，这样在 jsp 中就无法引入静态资源文件了。所以在 spring-mvc.xml 配置拦截器时需要进行优化，将资源文件排除在拦截器外。

```
<mvc:interceptors>
    <mvc:interceptor>
        <mvc:mapping path="/**/*"/>
        <mvc:exclude-mapping path="/**/fonts/*"/>
        <mvc:exclude-mapping path="/**/*.css"/>
        <mvc:exclude-mapping path="/**/*.js"/>
        <mvc:exclude-mapping path="/**/*.png"/>
        <mvc:exclude-mapping path="/**/*.gif"/>
        <mvc:exclude-mapping path="/**/*.jpg"/>
        <mvc:exclude-mapping path="/**/*.jpeg"/>
        <mvc:exclude-mapping path="/**/*login*"/>
        <mvc:exclude-mapping path="/**/*Login*"/>
        <bean class="com.demo.model.LoginIntercepter"></bean>
    </mvc:interceptor>
</mvc:interceptors>
```

为了演示 Ajax 与 Controller 之间的交互，需要在项目中引入 js 文件，这里在 webapp 下创建了 js 文件夹，将 jquery-3.3.1.min.js 放在该文件夹下。除了配置拦截器外，还需要在 spring-mvc.xml 中配置对静态资源文件的访问。

```
<!-- 对静态资源文件的访问 方案一 （二选一） -->
<mvc:default-servlet-handler />
<!-- 对静态资源文件的访问 方案二 （二选一） -->
<mvc:resources mapping="/images/**" location="/images/"
cache-period="31556926"/>
<mvc:resources mapping="/js/**" location="/js/"
cache-period="31556926"/>
<mvc:resources mapping="/css/**" location="/css/"
cache-period="31556926"/>
```

2. Ajax 发起请求

在 index.jsp 中增加了一个按钮，并为该按钮增加了一个单击事件，在单击事件中向服务端发起 post 请求。

```
<%@ page language="java" contentType="text/html; charset=utf-8"
    pageEncoding="utf-8"%>
<!DOCTYPE html PUBLIC "-//W3C//DTD HTML 4.01 Transitional//EN"
"http://www.w3.org/TR/html4/loose.dtd">
<html>
<head>
<meta http-equiv="Content-Type" content="text/html; charset=utf-8">
<script type="text/javascript" src="<%=request.getContextPath() %>
/js/jquery-3.3.1.min.js"></script>
<%
    String path = request.getContextPath();
    String basePath = request.getScheme() + "://"
            + request.getServerName() + ":" + request.getServerPort()
            + path + "/";
%>
<script type= "text/javascript" src= "<%=basePath %>js/
jquery-3.3.1.min.js"></script >
<script type="text/javascript">
$(document).ready(function(){
    $("#btnlogin").click(function(){
        var json = {
            'name':$(':input[name=name]').val(),
          'pwd':$(':input[name=pwd]').val()
          };
          var postdata = JSON.stringify(json);//json 对象转换 json 字符串
          $.ajax({
          type : 'POST',
          contentType : 'application/json;charset=UTF-8',//注意类型
```

```
                processData : false,
                url : '<%=path%>/index/requestbodybind.do',
                dataType : 'json',
                data : postdata,
                success : function(data) {
                    alert('user : '+data.name+'\npassword : '+data.pwd);
                },
                error : function(err) {
                    console.log(err.responseText);
                    alert(err.responseText);
                }
            });
        });
    });
    </script>
    <title>Insert title here</title>
    </head>
    <body>
    <form id="form1" name="myform" method="post" action="login.do" >
        用户:<input type="text" name="name"/>
        密码:<input type="password" name="pwd"/>
       <input type="button" value="登录" id="btnlogin"> <input
type="submit"/>
     </form>
    姓名:${user.name}<br>
    密码:${user.pwd}<br>
    </body>
    </html>
```

3. Controller 接收请求并响应

在 IndexController 中使用 requestBodyBind 方法来接收 User 类型对象，打印出该对象并返回。结果如图 7-20 所示。

图 7-20

```
        @RequestMapping(value="/requestbodybind.do",method =
{RequestMethod.POST})
        @ResponseBody
        public User requestBodyBind(@RequestBody User user){
            System.out.println("requestbodybind:" + user);
```

```
        return user;
    }
```

这里主要是有两个注解：@RequestBody 和@ResponseBody。

- @RequestBody 作用在形参列表上，用于将前台发送过来固定格式的数据【xml 格式或者 json 等】封装为对应的 JavaBean 对象，封装时使用到的一个对象是对系统默认配置的 HttpMessageConverter 进行解析，然后封装到形参上。
- @ResponseBody 是作用在方法上的，@ResponseBody 表示该方法的返回结果直接写入 HTTP response body 中，一般在异步获取数据时使用（也就是 AJAX），在使用 @RequestMapping 后，返回值通常解析为跳转路径，但是加上@ResponseBody 后返回结果不会被解析为跳转路径，而是直接写入 HTTP response body 中。比如异步获取 json 数据，加上@ResponseBody 后，会直接返回 json 数据。@RequestBody 将 HTTP 请求正文插入方法中，使用适合的 HttpMessageConverter 将请求体写入某个对象。上面在 requestBodyBind 方法中返回的 user 对象就没有被解析为跳转路径，而是直接写入 HTTP response body 中，在 ajax 响应结果中解析出来并弹出。

在实现的过程中可能会报 415 的错误，如图 7-21 所示。导致这个错误是由于在 pom.xml 中需要引入 json 转换相关类库，这里引入的是 jackson-databind 包。

HTTP Status 415 – Unsupported Media Type

Type Status Report

Description The origin server is refusing to service the request because the payload is in a format not supported by this method on the target resource.

图 7-21

```
<dependency>
      <groupId>com.fasterxml.jackson.core</groupId>
      <artifactId>jackson-databind</artifactId>
      <version>2.9.5</version>
   </dependency>
```

7.8 小结

本章主要介绍了 MVC 框架、Spring MVC 工作流程、view 与 Controller 之间的数据交互等内容。通过本章的学习，基本可以掌握 Spring MVC 框架的使用了。

第 8 章

◀ Spring Boot配置 ▶

本章学习 Spring 家族中一个全新的框架——Spring Boot。它可以用来简化 Spring 应用程序的创建和开发过程。

本章主要涉及的知识点：

- Spring Boot 引入：简单介绍、在线、离线安装。
- 基本配置：定制 Banner、配置文件。
- 外部配置：常规属性配置、基于 properties 类型安全的配置。
- 其他配置：Profile 配置、日志配置。
- 运行原理：运行原理介绍。

8.1 Spring Boot 基础

本节先了解一下 Spring Boot 基础知识，介绍它的优缺点，学习下它如何安装以及一些配置和运行原理。

8.1.1 Spring Boot 简介

Spring Boot 是由 Pivotal 团队提供的全新框架，其设计目的是用来简化新 Spring 应用的初始搭建以及开发过程。该框架使用了特定的方式来进行配置，从而使开发人员不再需要定义样板化的配置。通过这种方式，Boot 致力于在蓬勃发展的快速应用开发领域（rapid application development）成为领导者。

Spring Boot 有以下特点：

（1）能够快速创建基于 Spring 的应用程序。（简化配置）

（2）能够直接使用 Java 的 main 方法启动内嵌的 Tomcat，Jetty 服务器运行 Spring Boot 程序，不需要部署 war 包文件。

（3）提供约定的 starter POM 来简化 Maven 配置，让 Maven 配置变得简单。

（4）根据项目的 maven 依赖配置，Spring Boot 自动配置 Spring、SpringMVC 等其他开源框架。

（5）提供程序的健康检查等功能（检查内部的运行状态等）。

（6）基本可以完全不使用 xml 配置文件，采用注解配置。（或者默认约定的配置，代码中已经实现）

Spring Boot 可以支持快速地开发出 restful 风格的微服务架构，非常适合做微服务，单一 jar 包部署和管理也非常方便，不但配置精简，而且方便将 Spring 生态圈和其他工具链整合（比如 Redis、Email、Elasticsearch）。

8.1.2 在线安装

Eclipse 导入 Spring Boot 有两种方式：一种是在线安装，一种是离线安装。单击 Eclipse 中的 help→Eclipse Marketplace，搜索 sts，如图 8-1 所示，单击 Installed 按钮进行在线安装。

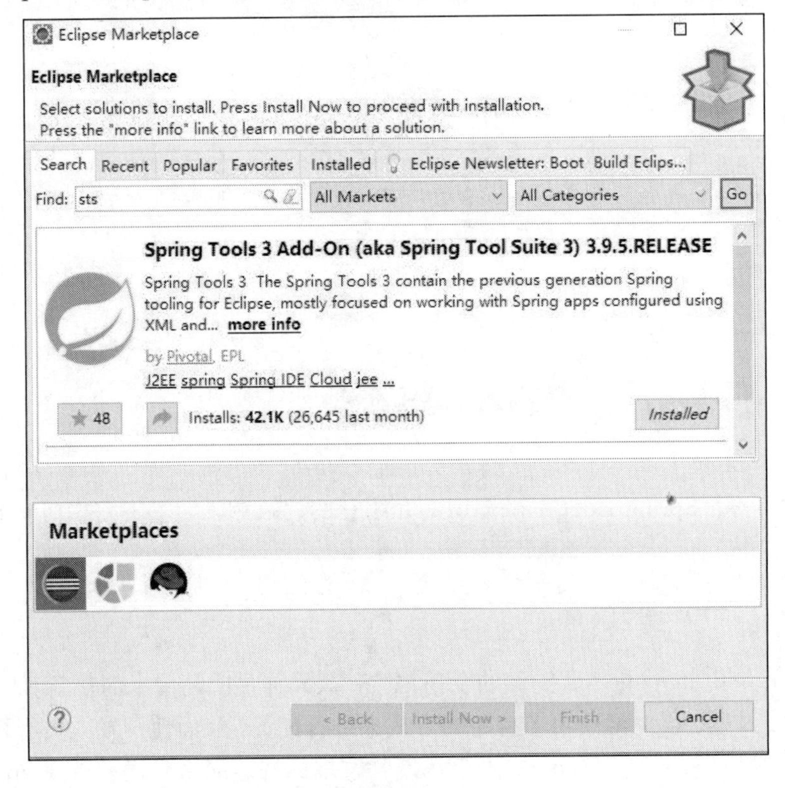

图 8-1

8.1.3 离线安装

离线安装时，可以先从 https://spring.io/tools3/sts/all 中下载 sts 插件，这里下载的是 springsource-tool-suite-3.9.4.RELEASE-e4.7.3a-updatesite 这个版本，然后选择 help→Install New Software→Add→Archive（由于这里没解压，因此选的是 Archive），选择要安装的组件，单击 Next，如图 8-2 所示。

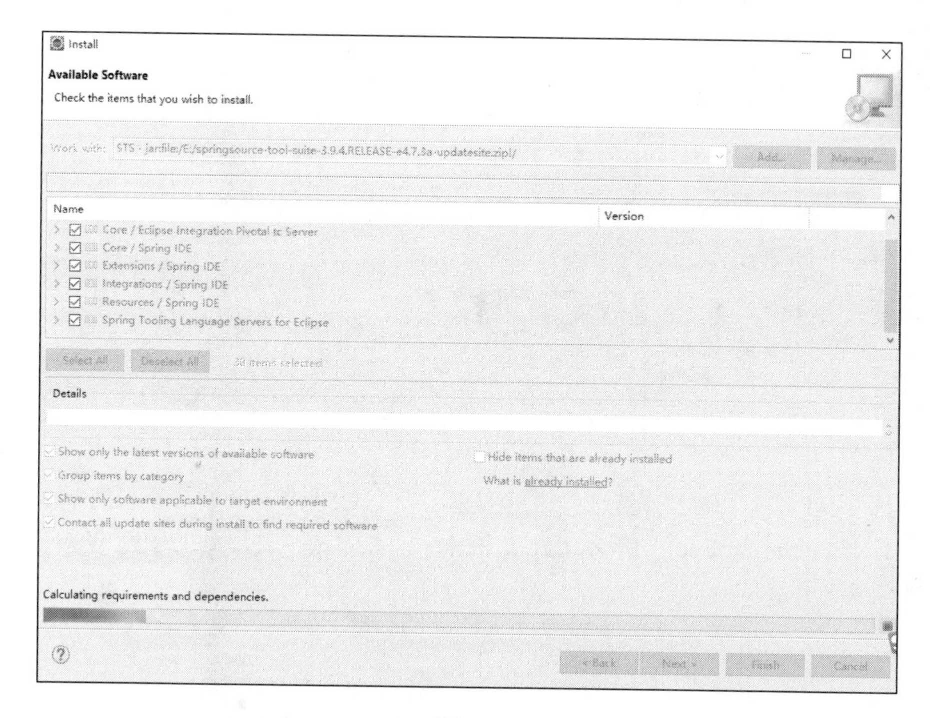

图 8-2

8.1.4 创建 Spring Boot 项目

安装 Spring Boot 插件之后，在 Eclipse 中创建 Project 时就会增加 Spring Boot 选项，如图 8-3 所示。

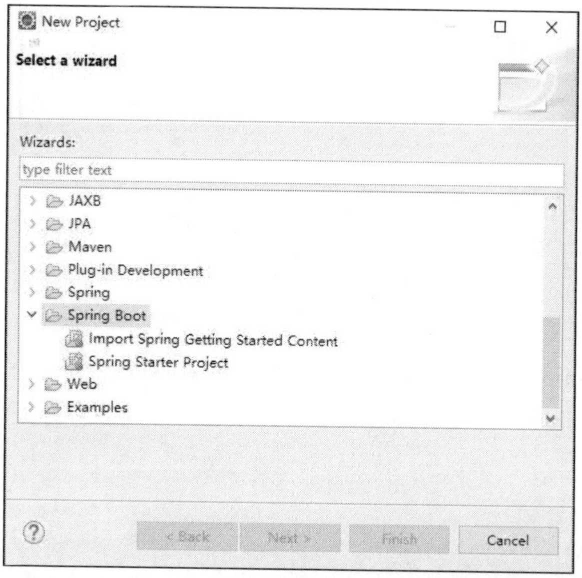

图 8-3

安装完之后我们创建一个简单的 Web 工程，了解一下 Spring Boot 项目创建的过程。首先在 File 菜单下单击 New→Project，选中 Spring Boot 下的 Spring Starter Project，如图 8-4 所示。

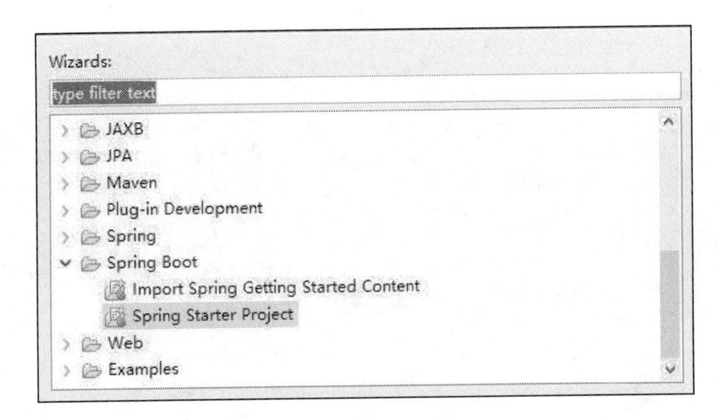

图 8-4

然后配置项目名称、版本等信息。这里创建了一个 HelloWorld 的项目来演示 Spring Boot 的简单使用。选择 Spring Boot 的版本（2.0.1 版本），并勾选项目用到的依赖 Web 模块，单击 Finish，如图 8-5 所示。

图 8-5

在 com.example.demo 包下新建 IndexController 类，在类中增加一个 Hello 方法，返回“Hello World”，如果@RestController 注解不理解也没关系，后续会有介绍。

```
package com.example.demo;

import org.springframework.web.bind.annotation.RequestMapping;
import org.springframework.web.bind.annotation.RestController;

@RestController
@RequestMapping("/index")
public class IndexController {
```

```
    @RequestMapping("/hello")
    public String Hello(){
        return "Hello World";
    }
}
```

选中项目，右击，依次选择 Run AS→Spring Boot App，一个简单的 Spring Boot 项目就运行成功了，在浏览器输入 url“http://localhost:8080/index/hello”之后，结果如图 8-6 所示。

图 8-6

8.2 Spring Boot 基本配置

本节先了解一下 Spring Boot 基础配置，主要包含定制 Banner、Spring Boot 配置文件、使用 xml 配置三部分。

8.2.1 定制 Banner

启动 Spring Boot 项目时，会在控制台打印默认的 banner，如果想炫酷一些，自己定义也是支持的，如图 8-7 所示。

图 8-7

首先在 src/main/resource 下新建 banner.txt，然后打开 http://patorjk.com/software/taag，输入要显示的文字，选择想要的样式，如图 8-8 所示。再复制到 banner.txt 中，再次启动时就会发现 banner 已变，如图 8-9 所示。

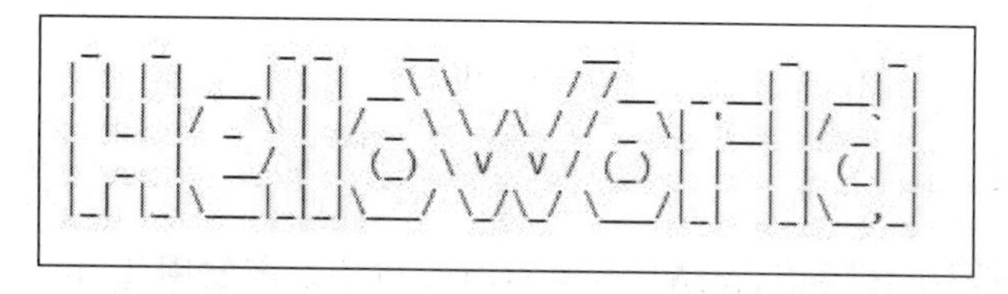

图 8-8

图 8-9

如果想关闭 banner，可以修改 main，设置 banner mode 为 OFF。

```
public static void main(String[] args) {
        //SpringApplication.run(HelloWorldApplication.class, args);
        SpringApplication app=new SpringApplication
(HelloWorldApplication.class);
        app.setBannerMode(Banner.Mode.OFF);
        app.run(args);
    }
```

8.2.2 配置文件

Spring Boot 使用一个全局的配置文件，配置文件名是固定的，application.properties 或 application.yml，可以用来修改 Spring Boot 自动配置的默认值。

这里在 application.properties 设置服务端的端口为 8081，再次启动项目时，服务器的端口已经变成了 8081，如图 8-10 所示。

```
server.port=8081
```

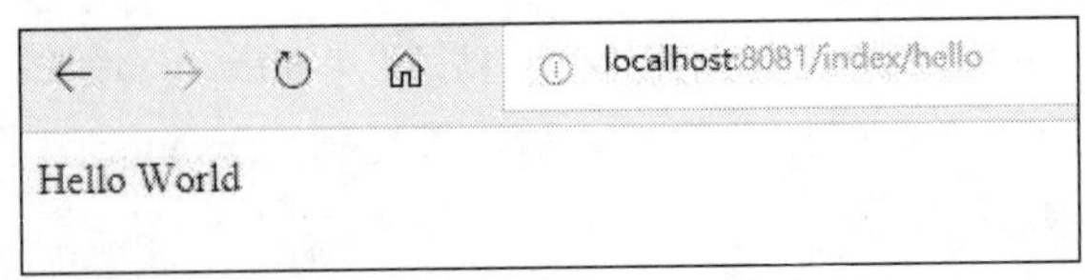

图 8-10

8.2.3 使用 xml 配置

虽然 Spring Boot 不提倡使用 xml 配置，但有时候也还是需要用的。这里我们在 src/main/java 下创建 com.example.services 包，在包下创建一个 bean 对象 HelloService，由于 Spring Boot

默认扫描的是 HelloWorldApplication main 方法对应的包以及子包，不会扫到 com.example.services 包，我们在 IndexController 注入一个该服务，然后启动，发现会报错，找不到该类。

定义 bean 对象 HelloService：

```
package com.example.services;
public class HelloService {

}
```

配置 bean 对象：

```
<?xml version="1.0" encoding="UTF-8"?>
<beans xmlns="http://www.springframework.org/schema/beans"
      xmlns:xsi="http://www.w3.org/2001/XMLSchema-instance"
      xsi:schemaLocation="http://www.springframework.org/schema/beans
      http://www.springframework.org/schema/beans/spring-beans.xsd">
  <bean id="helloService" class="com.example.services.HelloService">
  </bean>
</beans>
```

在 IndexController 注入 bean 对象：

```
package com.example.demo;
import org.springframework.beans.factory.annotation.Autowired;
import org.springframework.web.bind.annotation.RequestMapping;
import org.springframework.web.bind.annotation.RestController;
import com.example.services.HelloService;

@RestController
@RequestMapping("/index")
public class IndexController {

    @Autowired
    HelloService helloService;

    @RequestMapping("/hello")
    public String Hello(){
        return "Hello World";
    }
}
```

运行项目就会报下面的错误，找不到该 bean 对象，如图 8-11 所示。

```
    Field helloService in com.example.demo.IndexController required a bean of
type 'com.example.services.HelloService' that could not be found.
```

```
Description:

Field helloService in com.example.demo.IndexController required a bean of type 'com.example.services.HelloService' that could not be found.

The injection point has the following annotations:
	- @org.springframework.beans.factory.annotation.Autowired(required=true)


Action:

Consider defining a bean of type 'com.example.services.HelloService' in your configuration.
```

图 8-11

需要在 main 方法对应的包下创建配置文件引入 bean，以便让 Spring Boot 能扫描到。配置 ConfigClass 之后就可以成功启动项目。

```
package com.example.demo;

import org.springframework.context.annotation.Configuration;
import org.springframework.context.annotation.ImportResource;

@Configuration
@ImportResource(locations = {"application-bean.xml"})
public class ConfigClass {

}
```

8.3 Spring Boot 读取配置

在上一小节介绍了配置文件的使用，设置 server.port=8081 之后服务端启动的端口就变成了 8081，那它是怎么读取配置信息的呢？

8.3.1 读取核心配置文件

核心配置文件是指在 resources 根目录下的 application.properties 或 application.yml 配置文件。读取这两个配置文件的方法有以下两种，都比较简单。

1. 使用@Value 方式（常用）

在@Value 的${}中包含的是核心配置文件中的键名。在 IndexController 中增加 Name 属性，并使用@Value 为其赋值。在 hello()中同时返回 Name 的值，如图 8-12 所示。

```
@Value("${Test.Name}")
private String Name;

@RequestMapping("/hello")
public String Hello(){
    return "Hello World,"+Name;
}
```

图 8-12

2. 使用 Environment 方式

这种方式是依赖注入 Evnironment 来完成的，在创建的成员变量 private Environment env 上加上@Autowired 注解即可完成依赖注入，然后使用 env.getProperty("键名")即可读取出对应的值。结果如图 8-13 所示。

```
@Autowired
private Environment env;

@RequestMapping("/hello2")
public String Hello2(){
    return "Hello World,"+env.getProperty("Test.Name");
}
```

localhost:8081/index/hello2
Hello World,testName

图 8-13

8.3.2 读取自定义配置文件

为了不破坏核心文件的原生态，但又需要有自定义的配置信息存在，一般情况下会选择自定义配置文件来放这些自定义信息，这里在 resources 目录下创建配置文件 test.properties。Spring Boot 提供了类型安全的配置方式，通过@ConfigurationProperties 将 Properties 属性和一个 Bean 及其属性关联，从而实现类型安全的配置。

（1）在 src/main/resource 下创建一个 test.properties 的属性文件。

```
person.Name=xiaoming
person.Age=18
```

（2）创建 PersonSetting 类，用来与 test 属性文件进行关联。

在 com.example.demo 包下创建 PersonSetting 类，用来与 test 属性文件进行关联。

```
package com.example.demo;

import org.springframework.boot.context.properties.
ConfigurationProperties;
import org.springframework.context.annotation.PropertySource;
import org.springframework.stereotype.Component;
```

```
@Component
@PropertySource(value = "classpath:/test.properties")
@ConfigurationProperties(prefix="person")
public class PersonSetting {
   private String name;
   private int age;
   public String getName() {
      return name;
   }
   public void setName(String name) {
      this.name = name;
   }
   public int getAge() {
      return age;
   }
   public void setAge(int age) {
      this.age = age;
   }
}
```

（3）在 IndexController 注入 PersonSetting。

```
@Autowired
PersonSetting personSetting;
@RequestMapping("/hello3")
public String Hello3(){
   return "Hello World,Name:"+personSetting.getName()+",Age:"+
personSetting.getAge();
}
```

（4）启动，在浏览器中输入“http://localhost:8081/index/hello3”，可以显示出 PersonSetting 的属性值，如图 8-14 所示。

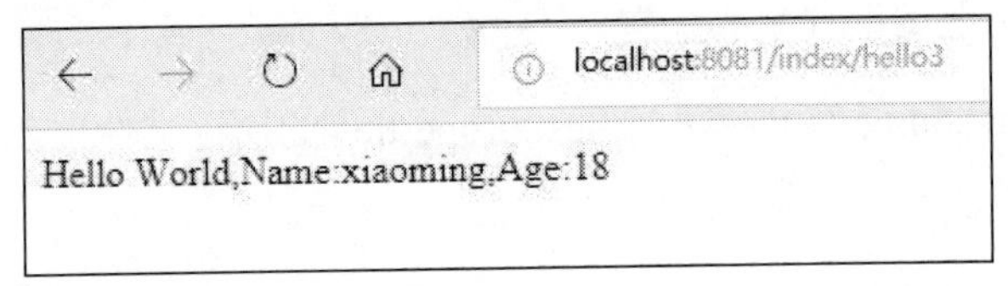

图 8-14

8.4 Profile 配置

在开发中可能会部署多个环境，每个环境部署的配置可能不一样。我们可以使用 application.properties 进行多个环境的配置，通过 application-{profile}.properties 来控制加载哪个环境的配置，将与环境无关的属性放置到 application.properties 文件里面，通过 spring.profiles.active=profiles 的值，加载不同环境的配置，如果不指定，就默认加载 application.properties 的配置，不会加载带有 profile 的配置。

我们可以创建 application-dev.properties 开发和 application-prod.properties 生产属性文件，分别指定不同的 port。在 application-dev.properties 文件中指定端口 8082，在 application-prod.properties 文件中指定端口 8083。

在 application.properties 中设置环境，如果设置的是 dev 环境，那么启动的端口就是 8082。如果设置的是 prod 环境，那么启动的端口就是 8083。这里设置的是 spring.profiles.active=prod，所以项目启动时端口号为 8083。结果如图 8-15 所示。

```
server.port=8081
Test.Name=testName
spring.profiles.active=prod
```

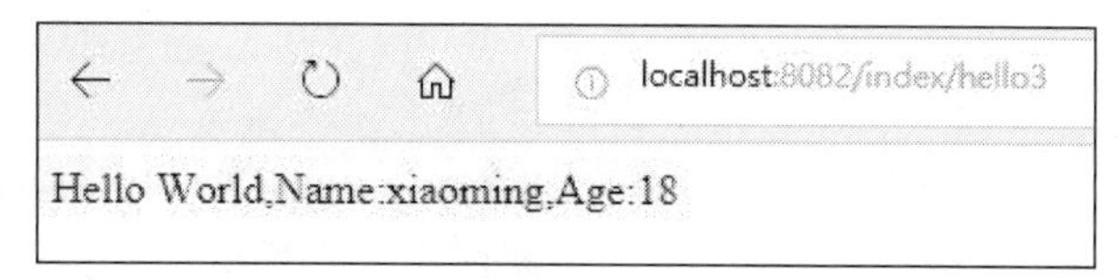

图 8-15

8.5 日志配置

8.5.1 简述

日志在项目中应该算是最重要的部分之一，尤其是上线之后，好的日志配置能够将其优势发挥到极致，大大降低后期的维护成本，好的日志配置能够让程序员一眼看出项目的问题，继而加以改善。这章我们就来学习一下 Spring Boot 中的日志配置。

Spring Boot 内部代码使用的是 commons-logging 来记录日志的，但是底层日志实现框架是可以随意替换的。Spring Boot 为 Java Util Logging、Log4J2 和 Logback 日志框架提供了默认配置，并且如果使用了 Starters，那么默认使用 Logback，如图 8-16 所示。

26. Logging

Spring Boot uses Commons Logging for all internal logging but leaves the underlying log implementation open. Default configurations are provided for Java Util Logging, Log4J2, and Logback. In each case, loggers are pre-configured to use console output with optional file output also available.

By default, if you use the "Starters", Logback is used for logging. Appropriate Logback routing is also included to ensure that dependent libraries that use Java Util Logging, Commons Logging, Log4J, or SLF4J all work correctly.

图 8-16

那么什么是 Starters 呢？其实只要你的 pom 文件中使用了 spring-boot-starter 就代表你使用了 Spring Boot 的 Starters。因为 Starters（启动器）是 Spring Boot 最核心的部件之一，没有了启动器，Spring Boot 就几乎废掉了。

8.5.2 Logback 的使用

Spring Boot 招人喜欢的一大特点就是配置方便，配置日志的相关参数也只需要写在 application.properties 中就可以了。当然，这仅仅是基本的配置，如果需要高级的配置，还是需要添加依赖所选择日志系统的配置文件。Spring Boot 的 Logging 配置的级别有 7 个：TRACE、DEBUG、INFO、WARN、ERROR、FATAL、OFF，如图 8-17 所示。

26.4 Log Levels

All the supported logging systems can have the logger levels set in the Spring `Environment` (for example, in `application.properties`) by using `logging.level.<logger-name>=<level>` where `level` is one of TRACE, DEBUG, INFO, WARN, ERROR, FATAL, or OFF. The `root` logger can be configured by using `logging.level.root`.

The following example shows potential logging settings in `application.properties`:

```
logging.level.root=WARN
logging.level.org.springframework.web=DEBUG
logging.level.org.hibernate=ERROR
```

图 8-17

在进行了这样的配置后，就可以在控制台打印 Log 信息了。但在生产环境中，日志往往要以文件形式存放到本地，那么 Spring Boot 的默认配置文件能够实现吗？答案是可以的，我们继续看官方文档，如图 8-18 所示。

26.3 File Output

By default, Spring Boot logs only to the console and does not write log files. If you want to write log files in addition to the console output, you need to set a `logging.file` or `logging.path` property (for example, in your `application.properties`).

The following table shows how the `logging.*` properties can be used together:

Table 26.1. Logging properties

`logging.file`	`logging.path`	Example	Description
(none)	*(none)*		Console only logging.
Specific file	*(none)*	`my.log`	Writes to the specified log file. Names can be an exact location or relative to the current directory.
(none)	Specific directory	`/var/log`	Writes `spring.log` to the specified directory. Names can be an exact location or relative to the current directory.

图 8-18

在默认情况下，Spring Boot 是仅仅在控制台打印 log 信息的，如果我们需要将 log 信息记录到文件，那么就需要在 application.properties 中配置 logging.file 或者 logging.path。配置 logging.file 的话是可以定位到自定义的文件的，使用 logging.path 的话，日志文件将使用 spring.log 来命名。而且日志文件会在 10Mb 大小的时候被截断，产生新的日志文件，默认级别为：ERROR、WARN、INFO。如果要自定义输出格式怎么办呢？其实在 application.properties 也是可以办到的。可以在 application.properties 中使用下面几个配置，不过部分只对默认的日志系统 Logback 起作用，如图 8-19 所示。

To help with the customization, some other properties are transferred from the Spring `Environment` to System properties, as described in the following table:

Spring Environment	System Property	Comments
`logging.exception-conversion-word`	`LOG_EXCEPTION_CONVERSION_WORD`	The conversion word used when logging exceptions.
`logging.file`	`LOG_FILE`	If defined, it is used in the default log configuration.
`logging.file.max-size`	`LOG_FILE_MAX_SIZE`	Maximum log file size (if LOG_FILE enabled). (Only supported with the default Logback setup.)
`logging.file.max-history`	`LOG_FILE_MAX_HISTORY`	Maximum number of archive log files to keep (if LOG_FILE enabled). (Only supported with the default Logback setup.)
`logging.path`	`LOG_PATH`	If defined, it is used in the default log configuration.
`logging.pattern.console`	`CONSOLE_LOG_PATTERN`	The log pattern to use on the console (stdout). (Only supported with the default Logback setup.)
`logging.pattern.dateformat`	`LOG_DATEFORMAT_PATTERN`	Appender pattern for log date format. (Only supported with the default Logback setup.)
`logging.pattern.file`	`FILE_LOG_PATTERN`	The log pattern to use in a file (if `LOG_FILE` is enabled). (Only supported with the default Logback setup.)
`logging.pattern.level`	`LOG_LEVEL_PATTERN`	The format to use when rendering the log level (default `%5p`). (Only supported with the default Logback setup.)
`PID`	`PID`	The current process ID (discovered if possible and when not already defined as an OS environment variable).

图 8-19

在 application.properties 中将环境切为开发模式，然后设置了日志的级别、日志文件存放位置、日志在控制台输出的格式、日志在日志文件输出的格式。

```
spring.profiles.active=dev
logging.level.root=INFO
logging.level.org.springframework.web=DEBUG
logging.file=E:/log/log.log
logging.pattern.console=%d{yyyy/MM/dd-HH:mm:ss} [%thread] %-5level %logger- %msg%n
logging.pattern.file=%d{yyyy/MM/dd-HH:mm} [%thread] %-5level %logger- %msg%n
```

再次启动项目，则会在日志文件路径 E:\log 下找到 log.log 的日志，如图 8-20 所示。

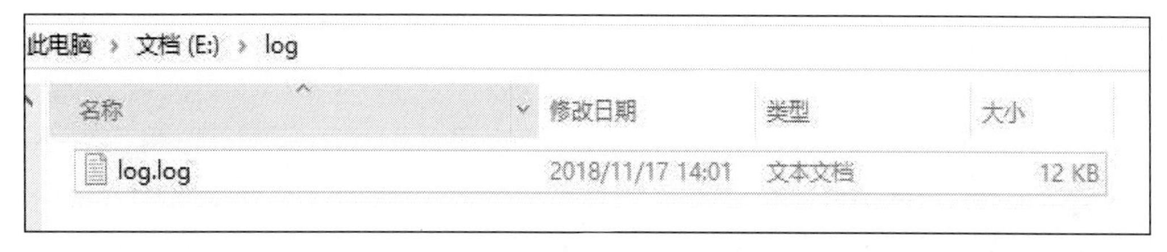

图 8-20

application.properties 中的配置有的时候不能满足我们的要求，或者我们要使用其他的集成进 Spring Boot 的日志系统，该怎么办呢？官方文档又给了我们答案，如图 8-21 所示。

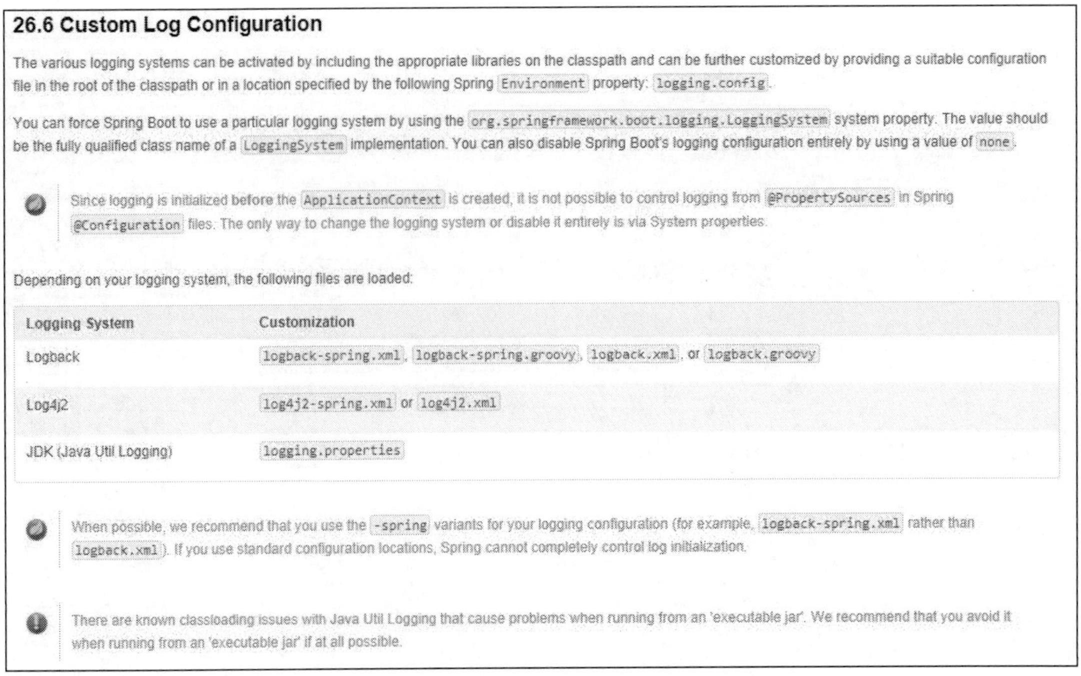

26.6 Custom Log Configuration

The various logging systems can be activated by including the appropriate libraries on the classpath and can be further customized by providing a suitable configuration file in the root of the classpath or in a location specified by the following Spring `Environment` property: `logging.config`.

You can force Spring Boot to use a particular logging system by using the `org.springframework.boot.logging.LoggingSystem` system property. The value should be the fully qualified class name of a `LoggingSystem` implementation. You can also disable Spring Boot's logging configuration entirely by using a value of `none`.

Since logging is initialized before the `ApplicationContext` is created, it is not possible to control logging from `@PropertySources` in Spring `@Configuration` files. The only way to change the logging system or disable it entirely is via System properties.

Depending on your logging system, the following files are loaded:

Logging System	Customization
Logback	`logback-spring.xml`, `logback-spring.groovy`, `logback.xml`, or `logback.groovy`
Log4j2	`log4j2-spring.xml` or `log4j2.xml`
JDK (Java Util Logging)	`logging.properties`

When possible, we recommend that you use the `-spring` variants for your logging configuration (for example, `logback-spring.xml` rather than `logback.xml`). If you use standard configuration locations, Spring cannot completely control log initialization.

There are known classloading issues with Java Util Logging that cause problems when running from an 'executable jar'. We recommend that you avoid it when running from an 'executable jar' if at all possible.

图 8-21

（1）通过将适当的库添加到 classpath，可以激活各种日志系统。然后在 classpath 的根目录（root）或通过 Spring Environment（application.properties）的 logging.config 属性指定的位置提供一个合适的配置文件来达到进一步的定制。（注意，由于日志是在 ApplicationContext 被创建之前初始化的，因此不可能在 Spring 的@Configuration 文件中通过@PropertySources 控制日志。系统属性和平常的 Spring Boot 外部配置文件能正常工作。）

（2）如果我们使用指定日志系统的配置文件，application.properties 中相关的日志配置是可以不要的。

（3）支持的三种日志系统（Logback，Log4j2（Log4j 也是支持的），JDKLogging）所识别的配置文件名。

使用 Logback 的指定配置文件实现更高级的日志配置，启用的方式很简单，在 classpath 的 resources 下新建 logback.xml 文件即可。

```
<?xml version="1.0" encoding="UTF-8"?>
<configuration>
  <!-- 控制台打印日志的相关配置 -->
  <appender name="STDOUT" class="ch.qos.logback.core.ConsoleAppender">
    <!-- 日志格式 -->
    <encoder>
        <pattern>%d{yyyy-MM-dd HH:mm:ss} [%level] - %m%n</pattern>
    </encoder>
    <!-- 日志级别过滤器 -->
    <filter class="ch.qos.logback.classic.filter.LevelFilter">
      <!-- 过滤的级别 -->
```

```
        <level>DEBUG</level>
        <!-- 匹配时的操作：接收（记录） -->
        <onMatch>ACCEPT</onMatch>
        <!-- 不匹配时的操作：拒绝（不记录） -->
        <onMismatch>DENY</onMismatch>
      </filter>
    </appender>

    <!-- 文件保存日志的相关配置 -->
    <appender name="ERROR-OUT" class="ch.qos.logback.core.rolling.
RollingFileAppender">
       <!-- 保存日志文件的路径 -->
      <file>E:/log/log.log</file>
      <!-- 日志格式 -->
      <encoder>
         <pattern>%d{yyyy-MM-dd HH:mm:ss} [%class:%line] - %m%n</pattern>
      </encoder>
      <!-- 日志级别过滤器 -->
      <filter class="ch.qos.logback.classic.filter.LevelFilter">
        <!-- 过滤的级别 -->
        <level>DEBUG</level>
        <!-- 匹配时的操作：接收（记录） -->
        <onMatch>ACCEPT</onMatch>
        <!-- 不匹配时的操作：拒绝（不记录） -->
        <onMismatch>DENY</onMismatch>
      </filter>
      <!-- 循环政策：基于时间创建日志文件 -->
      <rollingPolicy class="ch.qos.logback.core.rolling.
TimeBasedRollingPolicy">
        <!-- 日志文件名格式 -->
        <fileNamePattern>error.%d{yyyy-MM-dd}.log</fileNamePattern>
        <!-- 最大保存时间：30 天-->
        <maxHistory>30</maxHistory>
      </rollingPolicy>
    </appender>

    <!-- 基于 dubug 处理日志：具体控制台或者文件对日志级别的处理还要看所在 appender 配置
的 filter，如果没有配置 filter，则使用 root 配置 -->
    <root level="debug">
      <appender-ref ref="STDOUT" />
      <appender-ref ref="ERROR-OUT" />
    </root>
  </configuration>
```

8.5.3　Log4j2 的使用

在创建 Spring Boot 工程时，我们引入了 spring-boot-starter，其中包含了 spring-boot-starter-logging，该依赖内容就是 Spring Boot 默认的日志框架 Logback，所以我们在引入 Log4j2 之前，需要先排除该包的依赖再引入 Log4j2 的依赖。

```
<dependency>
      <groupId>org.springframework.boot</groupId>
      <artifactId>spring-boot-starter-web</artifactId>
      <exclusions>
         <exclusion>
            <groupId>org.springframework.boot</groupId>
            <artifactId>spring-boot-starter-logging</artifactId>
         </exclusion>
      </exclusions>
</dependency>
<dependency>
       <groupId>org.springframework.boot</groupId>
      <artifactId>spring-boot-starter-log4j2</artifactId>
</dependency>
```

和 Logback 一样，我们可以使用 application.properties，如果需要更高级的配置选择，必须要添加 Log4j2 的配置文件了。我们在 classpath 的 resources 下新建 log4j2.xml 文件：

```
<?xml version="1.0" encoding="UTF-8"?>
<configuration>
   <appenders>
      <Console name="Console" target="SYSTEM_OUT">
         <ThresholdFilter level="trace" onMatch="ACCEPT"
onMismatch="DENY" />
         <PatternLayout pattern="%d{HH:mm:ss.SSS} %-5level %class{36} %L
%M - %msg%xEx%n" />
      </Console>
      <File name="log" fileName="E:/log/test.log" append="false">
         <PatternLayout pattern="%d{HH:mm:ss.SSS} %-5level %class{36} %L
%M - %msg%xEx%n" />
      </File>
      <RollingFile name="RollingFile" fileName="E:/log/spring.log"
filePattern="log/$${date:yyyy-MM}/app-%d{MM-dd-yyyy}-%i.log">
         <PatternLayout pattern="%d{yyyy-MM-dd 'at' HH:mm:ss z} %-5level
%class{36} %L %M - %msg%xEx%n" />
         <SizeBasedTriggeringPolicy size="50MB" />
      </RollingFile>
   </appenders>
   <loggers>
      <root level="DEBUG">
         <appender-ref ref="RollingFile" />
         <appender-ref ref="Console" />
      </root>
   </loggers>
```

```
</configuration>
```

重新启动项目，在 E:\log 文件夹下就能看到 3 个日志文件，如图 8-22 所示。

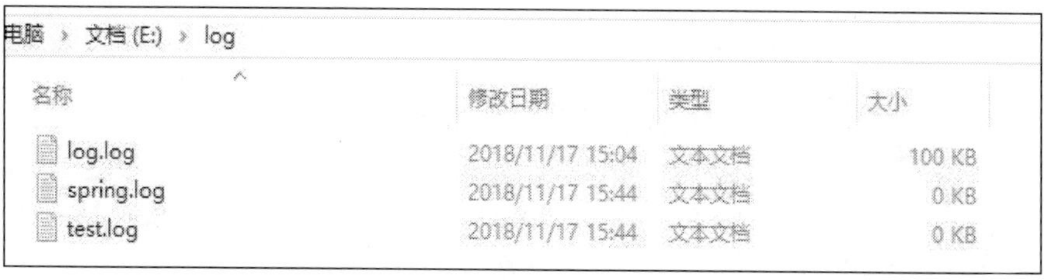

电脑 › 文档 (E:) › log

名称	修改日期	类型	大小
log.log	2018/11/17 15:04	文本文档	100 KB
spring.log	2018/11/17 15:44	文本文档	0 KB
test.log	2018/11/17 15:44	文本文档	0 KB

图 8-22

8.6 运行原理

8.6.1 习惯优于配置

Spring Boot 出现之后，得益于“习惯优于配置”这个理念，再也没有烦琐的配置、难以集成的内容（大多数流行第三方技术都被集成在内）。 那么背后实现的核心原理到底是什么呢？Spring Boot 关于自动配置的源码在 spring-boot-autoconfigure-x.x.x.x.jar 中，主要包含了如图 8-23 所示的配置。

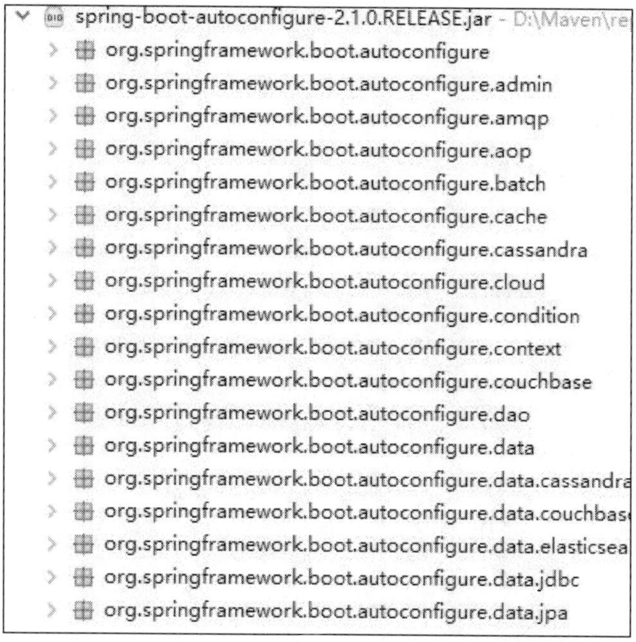

spring-boot-autoconfigure-2.1.0.RELEASE.jar - D:\Maven\re
- org.springframework.boot.autoconfigure
- org.springframework.boot.autoconfigure.admin
- org.springframework.boot.autoconfigure.amqp
- org.springframework.boot.autoconfigure.aop
- org.springframework.boot.autoconfigure.batch
- org.springframework.boot.autoconfigure.cache
- org.springframework.boot.autoconfigure.cassandra
- org.springframework.boot.autoconfigure.cloud
- org.springframework.boot.autoconfigure.condition
- org.springframework.boot.autoconfigure.context
- org.springframework.boot.autoconfigure.couchbase
- org.springframework.boot.autoconfigure.dao
- org.springframework.boot.autoconfigure.data
- org.springframework.boot.autoconfigure.data.cassandra
- org.springframework.boot.autoconfigure.data.couchbas
- org.springframework.boot.autoconfigure.data.elasticsea
- org.springframework.boot.autoconfigure.data.jdbc
- org.springframework.boot.autoconfigure.data.jpa

图 8-23

我们可以在这里看见所有 Spring Boot 为我们做的自动配置。通过在 application.properties 中设置属性“debug=true”，可以通过控制台的输出观察自动配置启动的情况，如图 8-24 所示。

```
Positive matches:
-----------------

   CodecsAutoConfiguration matched:
      - @ConditionalOnClass found required class 'org.springframework.http.codec.CodecConfigurer' (OnClassCondition)

   CodecsAutoConfiguration.JacksonCodecConfiguration matched:
      - @ConditionalOnClass found required class 'com.fasterxml.jackson.databind.ObjectMapper' (OnClassCondition)

   CodecsAutoConfiguration.JacksonCodecConfiguration#jacksonCodecCustomizer matched:
      - @ConditionalOnBean (types: com.fasterxml.jackson.databind.ObjectMapper; SearchStrategy: all) found bean 'jacksonObjectMappe

   DispatcherServletAutoConfiguration matched:
      - @ConditionalOnClass found required class 'org.springframework.web.servlet.DispatcherServlet' (OnClassCondition)
      - found ConfigurableWebEnvironment (OnWebApplicationCondition)
```

图 8-24

8.6.2 运行原理

在第一次使用 Spring Boot 的时候，大家都会惊讶于@SpringBootApplication 这个注解，有了它马上就能够让整个应用跑起来。实际上它只是一个组合注解，包含@Configuration、@EnableAutoConfiguration、@ComponentScan 这三个注解。

```
@Target(ElementType.TYPE)
@Retention(RetentionPolicy.RUNTIME)
@Documented
@Inherited
@SpringBootConfiguration
@EnableAutoConfiguration
@ComponentScan(excludeFilters = {
        @Filter(type = FilterType.CUSTOM, classes = TypeExcludeFilter. class),
        @Filter(type = FilterType.CUSTOM, classes =
AutoConfigurationExcludeFilter.class) })
public @interface SpringBootApplication {

    @AliasFor(annotation = EnableAutoConfiguration.class)
    Class<?>[] exclude() default {};

    @AliasFor(annotation = EnableAutoConfiguration.class)
    String[] excludeName() default {};

    @AliasFor(annotation = ComponentScan.class, attribute = "basePackages")
    String[] scanBasePackages() default {};

    @AliasFor(annotation = ComponentScan.class, attribute =
"basePackageClasses")
    Class<?>[] scanBasePackageClasses() default {};

}
```

它的核心功能是由@EnableAutoConfiguration 这个注解提供的，我们来看看@EnableAutoConfiguration 的源代码：

```
@Target(ElementType.TYPE)
@Retention(RetentionPolicy.RUNTIME)
@Documented
```

```
@Inherited
@AutoConfigurationPackage
@Import(AutoConfigurationImportSelector.class)
public @interface EnableAutoConfiguration {

    String ENABLED_OVERRIDE_PROPERTY =
"spring.boot.enableautoconfiguration";

    Class<?>[] exclude() default {};

    String[] excludeName() default {};

}
```

这里的关键功能是@Import 注解导入的配置功能 EnableAutoConfigurationImportSelector 使用 SpringFactoriesLoader.loadFactoryNames 方法来扫描具有 META-INF/spring.factories 文件的 jar 包，spring-boot-autoconfigure-x.x.x.x.jar 里就有一个 spring.factories 文件，这个文件中声明了有哪些要自动配置。

下面我们来分析一下 spring boot autoconfigure 里面的 JdbcTemplateAutoConfiguration，应该就会明白这套自动配置机制到底是怎么一回事了：

```
@Configuration
@ConditionalOnClass({ DataSource.class, JdbcTemplate.class })
@ConditionalOnSingleCandidate(DataSource.class)
@AutoConfigureAfter(DataSourceAutoConfiguration.class)
@EnableConfigurationProperties(JdbcProperties.class)
public class JdbcTemplateAutoConfiguration {

    @Configuration
    static class JdbcTemplateConfiguration {

        private final DataSource dataSource;

        private final JdbcProperties properties;

        JdbcTemplateConfiguration(DataSource dataSource, JdbcProperties
properties) {
            this.dataSource = dataSource;
            this.properties = properties;
        }

        @Bean
        @Primary
        @ConditionalOnMissingBean(JdbcOperations.class)
        public JdbcTemplate jdbcTemplate() {
            JdbcTemplate jdbcTemplate = new JdbcTemplate(this.dataSource);
            JdbcProperties.Template template =
this.properties.getTemplate();
            jdbcTemplate.setFetchSize(template.getFetchSize());
            jdbcTemplate.setMaxRows(template.getMaxRows());
            if (template.getQueryTimeout() != null) {
```

```
                jdbcTemplate
                    .setQueryTimeout((int) template.getQueryTimeout().
getSeconds());
            }
            return jdbcTemplate;
        }

    }

    @Configuration
    @Import(JdbcTemplateConfiguration.class)
    static class NamedParameterJdbcTemplateConfiguration {

        @Bean
        @Primary
        @ConditionalOnSingleCandidate(JdbcTemplate.class)
        @ConditionalOnMissingBean(NamedParameterJdbcOperations.class)
        public NamedParameterJdbcTemplate namedParameterJdbcTemplate(
                JdbcTemplate jdbcTemplate) {
            return new NamedParameterJdbcTemplate(jdbcTemplate);
        }

    }

}
```

首先这被@Configuration 注解了，是一个配置类，当满足以下条件时这个 bean 被装配：

（1）DataSource、DataSource 在类路径下。

（2）DataSource 指定的类在 BeanFactory 中只有一个候选的 bean，或者有多个候选的 bean，但是其中一个指定了 primary。

（3）在指定的配置类 DataSourceAutoConfiguration 初始化后再加载。

我们可以看一下通过@EnableConfigurationProperties(JdbcProperties.class)自动注入的属性（这是习惯优于配置的最终落地点）：

```
@ConfigurationProperties(prefix = "spring.jdbc")
public class JdbcProperties {

    private final Template template = new Template();

    public Template getTemplate() {
        return this.template;
    }

    public static class Template {

        private int fetchSize = -1;

        private int maxRows = -1;

        @DurationUnit(ChronoUnit.SECONDS)
        private Duration queryTimeout;
```

```
        public int getFetchSize() {
            return this.fetchSize;
        }

        public void setFetchSize(int fetchSize) {
            this.fetchSize = fetchSize;
        }

        public int getMaxRows() {
            return this.maxRows;
        }

        public void setMaxRows(int maxRows) {
            this.maxRows = maxRows;
        }

        public Duration getQueryTimeout() {
            return this.queryTimeout;
        }

        public void setQueryTimeout(Duration queryTimeout) {
            this.queryTimeout = queryTimeout;
        }

    }

}
```

在我们什么都不干的情况下，只需要引入数据库相关依赖，用 JdbcTemplate 访问数据库即可。同时我们通过在 application.properties 中修改 spring.jdbc 相关的参数就能够修改连接配置。利用这个规则，我们也可以轻松地把目前 Spring Boot 还未集成的、我们自己要使用的第三方技术自动集成起来。

8.7 小结

本节主要学习 Spring Boot 的基础知识，介绍了 Spring Boot 插件的安装、基本配置、Profile 配置、日志配置以及运行原理，为后续 Spring Boot 的应用打下基础。

第 9 章

Spring Boot的应用

上一章学习了 Spring Boot 的环境安装、基本配置等内容，不但配置精简，而且方便将 Spring 生态圈和其他工具链整合，本章将了解一下 Spring Boot 的具体应用。

本章主要涉及的知识点：

- Spring Boot 集成 Web：集成 Thymeleaf、JSP。
- Spring Boot 集成 Data：集成 MyBatis、Redis、RabbitMQ。
- Spring Boot 集成其他工具：集成 Druid、打包测试部署、定时任务、邮件发送。

9.1 Spring Boot 之 Web

Web 开发都会涉及前端页面展示，在以往的 Spring MVC 框架中一般使用 JSP 来显示页面，那 Spring Boot 框架中又是用什么显示 Web 呢？

9.1.1 Spring Boot 集成 Thymeleaf

Spring Boot 提供了大量的模板引擎，包含了 FreeMarker、Groovy、Thymeleaf、Velocity 和 Mustache。Spring Boot 中推荐使用 Thymeleaf 作为模板引擎，因为 Thymeleaf 提供了完美的 Spring MVC 的支持。Thymeleaf 是一个 Java 类库，它是一个 xml/xhtml/html5 的模板引擎，可以作为 MVC 的 Web 应用的 View 层。Thymeleaf 还提供了额外的模块与 Spring MVC 集成，所以我们可以使用 Thymeleaf 完全替代 JSP。我们来看一下 Thymeleaf 的简单使用。

1. Thymeleaf 配置

Thymeleaf 有哪些属性可以配置呢？我们可以在 org.springframework.boot.autoconfigure.thymeleaf 下的 ThymeleafProperties.class，如图 9-1 所示。

org.springframework.boot.autoconfigure.thymeleaf
ThymeleafAutoConfiguration.class
ThymeleafProperties.class
ThymeleafTemplateAvailabilityProvider.class

图 9-1

由于 Spring Boot 约定大于配置，因此在 ThymeleafProperties.class 中也都有默认值，如果我们想改变默认值，可以在 application.properties 设置，属性的前缀为 spring.thymeleaf。这里用的都是它的默认值。默认路径在 templates 下，文件是 html 文件，如图 9-2 所示。

```
@ConfigurationProperties(prefix = "spring.thymeleaf")
public class ThymeleafProperties {

    private static final Charset DEFAULT_ENCODING = StandardCharsets.UTF_8;

    public static final String DEFAULT_PREFIX = "classpath:/templates/";

    public static final String DEFAULT_SUFFIX = ".html";

    /**
     * Whether to check that the template exists before rendering it.
     */
    private boolean checkTemplate = true;

    /**
     * Whether to check that the templates location exists.
     */
    private boolean checkTemplateLocation = true;

    /**
     * Prefix that gets prepended to view names when building a URL.
     */
    private String prefix = DEFAULT_PREFIX;

    /**
     * Suffix that gets appended to view names when building a URL.
     */
    private String suffix = DEFAULT_SUFFIX;

    /**
     * Template mode to be applied to templates. See also Thymeleaf's TemplateMode enum.
     */
```

图 9-2

2. 项目引入 Thymeleaf

这里还是在 HelloWorld 的例子基础上进行修改，需要在 pom.xml 中引入 Thymeleaf。注意一下，由于用的是 spring5，如果引入的 Thymeleaf 版本不正确就可能会报错，而且不同的 spring 引入 Thymeleaf 的 artifactId 也不一样。

```
<dependency>
      <groupId>org.thymeleaf</groupId>
      <artifactId>thymeleaf-spring5</artifactId>
      <version>3.0.9.RELEASE</version>
</dependency>
```

3. 测试

这里创建一个 User 类，并增加 getUsers()方法来返回 user 列表。

```
package com.example.demo;

import java.util.ArrayList;
import java.util.List;

public class User {

   private String name;
```

```
    private int age;

    public int getAge() {
        return age;
    }
    public void setAge(int age) {
        this.age = age;
    }
    public String getName() {
        return name;
    }
    public void setName(String name) {
        this.name = name;
    }

    public User(String name, int age) {
        super();
        this.name = name;
        this.age = age;
    }

    public static List<User>getUsers()
    {
        List<User> users=new ArrayList<User>();
        for(int i=0;i<5;i++)
        {
            User user=new User("小明"+i,25+i);
            users.add(user);
        }
        return users;
    }
}
```

新建 HelloWorldController，用 hello 方法将 user 列表加入 model 的属性中。

```
package com.example.demo;

import org.springframework.stereotype.Controller;
import org.springframework.ui.Model;
import org.springframework.web.bind.annotation.RequestMapping;
import org.springframework.web.bind.annotation.RequestMethod;


@Controller
@RequestMapping("/helloworld")
public class HelloWorldController {
    @RequestMapping(value = "/hello4",method = RequestMethod.GET)
```

```
    public String hello(Model model) {
        model.addAttribute("users", User.getUsers());
        return "hello";
    }
}
```

然后在 classpath 路径下的 templates 目录下新建 hello.html 文件。这里要注意，需要在 html 中加入下面一行代码：<html xmlns:th="http://www.thymeleaf.org"> 。

```
<!DOCTYPE HTML>
<html xmlns:th="http://www.thymeleaf.org">
<head>
    <title>hello</title>
    <meta http-equiv="Content-Type" content="text/html; charset=UTF-8" />
</head>
<body>
  <table>
      <thead>
          <tr>
             <th>序号</th>
              <th>姓名</th>
              <th>年龄</th>
          </tr>
      </thead>
      <tbody>
          <tr th:each="user,userState : ${users}">
              <td th:text="${userState.index}"></td>
              <td th:text="${user.name}"></td>
              <td th:text="${user.age}"></td>
          </tr>
      </tbody>
  </table>
</body>
</html>
```

在 hello.html 中使用 th:each 标签来迭代循环，语法：th:each="obj,iterStat:${objList}"。迭代对象可以是 java.util.List、java.util.Map、数组等。这里只列举一下 th:each 标签的使用，关于 Thymeleaf 的使用，可以参考它的官方文档，这里不详细介绍。

iterStat 称作状态变量，属性有：

- Index：当前迭代对象的 index（从 0 开始计算）。
- Count：当前迭代对象的 index（从 1 开始计算）。
- Size：被迭代对象的大小。
- Current：当前迭代变量。
- even/odd：布尔值，当前循环是否是偶数/奇数（从 0 开始计算）。
- First：布尔值，当前循环是否是第一个。

- Last：布尔值，当前循环是否是最后一个。

最后启动项目，在浏览器输入 URL“http://localhost:8081/helloworld/hello4”，之后就会在浏览器输出 user 列表内容，如图 9-3 所示。

图 9-3

9.1.2 Spring Boot 集成 JSP

虽然 Spring Boot 推荐使用 Thymeleaf 作为模板引擎，也是支持使用 JSP 的，一些旧项目在转 Spring Boot 时也会方便一些，本小节就来学习一下 Spring Boot 集成 JSP 页面。

这里还是在 HelloWorld 项目基础上进行修改，这次是集成 JSP，所以要先引入 JSP 的依赖，同时把 Thymeleaf 的依赖去掉。

```
<dependency>
        <groupId>org.apache.tomcat.embed</groupId>
        <artifactId>tomcat-embed-jasper</artifactId>
        <scope>provided</scope>
</dependency>
<dependency>
        <groupId>javax.servlet</groupId>
        <artifactId>jstl</artifactId>
        <scope>provided</scope>
</dependency>
<dependency>
        <groupId>javax.servlet</groupId>
        <artifactId>javax.servlet-api</artifactId>
        <scope>provided</scope>
</dependency>
```

同时，集成 JSP，肯定少不了 JSP 页面，这里把 JSP 页面 hello.jsp 放在了 /demo/src/main/webapp/view 下，在 hello.jsp 页面还是遍历集成 Thymeleaf 时用的 users。

```
<%@ page language="java" contentType="text/html; charset=utf-8"
   pageEncoding="utf-8"%>
<%@ taglib prefix="c" uri="http://java.sun.com/jsp/jstl/core"%>
<!DOCTYPE html PUBLIC "-//W3C//DTD HTML 4.01 Transitional//EN"
"http://www.w3.org/TR/html4/loose.dtd">
<html>
<head>
```

```
<meta http-equiv="Content-Type" content="text/html; charset=utf-8">
<title>Insert title here</title>
</head>
<body>
  <table border="1">
        <thead>
            <tr>
               <th>序号</th>
                <th>姓名</th>
                <th>年龄</th>
            </tr>
        </thead>
            <c:forEach var="data" items="${users}" varStatus="loop">
                <tr>
                    <td>${loop.index + 1}</td>
                    <td>${data.name}</td>
                    <td>${data.age}</td>
                </tr>
            </c:forEach>
  </table>
</body>
</html>
```

然后还需要在 application.properties 中设置 view 的映射关系。

```
spring.mvc.view.prefix=/view/
spring.mvc.view.suffix=.jsp
```

最后在浏览器中输入 URL “http://localhost:8081/helloworld/hello4”，也会将 users 列表显示到页面中，如图 9-4 所示。

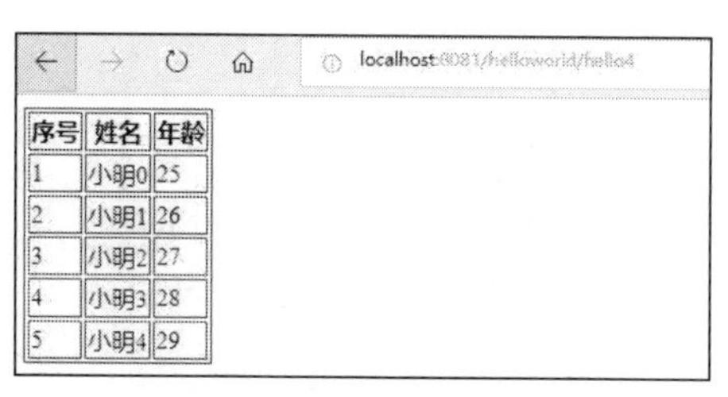

图 9-4

9.2 Spring Boot 之 Data

在上一节介绍了 Spring Boot 集成 Web 层，其中页面显示的数据是在 User 类中创建的，但通常页面显示的数据是从数据库或者其他地方获取的，本节就来介绍 Spring Boot 集成 Data 层操作，主要包含 Spring Boot 集成 MyBatis、Redis、RabbitMQ 这 3 种。

9.2.1 Spring Boot 集成 MyBatis

在上节中通过 getUsers()方法来获取用户列表显示在页面中，如果将 user 数据放在数据库中使用 MyBatis 来获取又是怎么操作的呢？我们也通过这个例子来了解 Spring Boot 集成 MyBatis 方法。Spring Boot 集成 MyBatis 一般有两种方式：一个是基于注解的，一个是基于 xml 配置的。

1. 基于注解的 MyBatis 集成

（1）引入依赖。因为是集成 MyBatis，肯定是要引入 MyBatis 相关的依赖 mybatis-spring-boot-starter，同时用的是 MySQL，所以也需要引入 MySQL 相关的 mysql-connector-java 依赖。

```
<dependency>
    <groupId>org.mybatis.spring.boot</groupId>
    <artifactId>mybatis-spring-boot-starter</artifactId>
    <version>1.3.2</version>
</dependency>
<dependency>
    <groupId>mysql</groupId>
    <artifactId>mysql-connector-java</artifactId>
    <version>8.0.11</version>
</dependency>
```

（2）在 MySQL 中创建一个名为 mybatis 的数据库，在里面创建一个 user 的表。在 User 类基础上增加了一个 id 属性，这样方便与数据库的表对照。

user 表：

```
CREATE TABLE `user` (
  `id` int(11) NOT NULL AUTO_INCREMENT,
  `name` varchar(20) DEFAULT NULL,
  `age` int(11) DEFAULT NULL,
  PRIMARY KEY (`id`)
) ENGINE=InnoDB DEFAULT CHARSET=utf8;
```

User 类：

```
package com.example.demo;

public class User {

    private int id;

    private String name;

    private int age;
```

```
    public int getId() {
        return id;
    }
    public void setId(int id) {
        this.id = id;
    }
    public int getAge() {
        return age;
    }
    public void setAge(int age) {
        this.age = age;
    }
    public String getName() {
        return name;
    }
    public void setName(String name) {
        this.name = name;
    }
}
```

（3）把 model 与操作数据库的 SQL 对照起来，用什么对照呢？创建一个 mapper，用@Insert、@Delete、@Update、@Select 注解分别对应 MyBatis 中的 select、delete、update、select 元素来实现 User 类与 user 表的关联。

@Results 注解来映射查询结果集到实体类属性，当数据库字段名与实体类对应的属性名不一致时，可以使用@Results 映射来将其对应起来。column 为数据库字段名，porperty 为实体类属性名，jdbcType 为数据库字段数据类型，id 为是否为主键。

当这段@Results 代码需要在多个方法用到时，为了提高代码复用性，我们可以为这个@Results 注解设置 id，然后使用@ResultMap 注解来复用这段代码。

```
package com.example.mapper;

import java.util.List;
import org.apache.ibatis.annotations.Delete;
import org.apache.ibatis.annotations.Insert;
import org.apache.ibatis.annotations.Result;
import org.apache.ibatis.annotations.ResultMap;
import org.apache.ibatis.annotations.Results;
import org.apache.ibatis.annotations.Select;
import org.apache.ibatis.annotations.Update;
import org.apache.ibatis.type.JdbcType;

import com.example.demo.User;

public interface UserMapper {

  @Select("SELECT * FROM user")
```

```
    @Results(id="userMap",value={

        @Result(property = "id",column = "id",jdbcType=JdbcType.INTEGER,
id=true),

        @Result(property = "age",column = "age",jdbcType=JdbcType.INTEGER),

        @Result(property = "name",column = "name",jdbcType=JdbcType.VARCHAR)

    })
    List<User> getAll();

    @Select("SELECT * FROM user WHERE id = #{id}")
    @ResultMap(value="userMap")
    User getOne(int id);

    @Insert("INSERT INTO user(name,age) VALUES(#{name}, #{age})")
    void insert(User user);

    @Update("UPDATE user SET name=#{userName},age=#{age} WHERE id =#{id}")
    void update(User user);

    @Delete("DELETE FROM user WHERE id =#{id}")
    void delete(int id);
}
```

（4）配置扫描。上面配置了 mapper，那怎么让系统知道 mapper 放在哪里呢?于是有了@MapperScan 注解。在 HelloWorldApplication 中使用@MapperScan 将 com.example.mapper 包中的映射文件配置进项目中。

```
@MapperScan("com.example.mapper")
@SpringBootApplication
public class HelloWorldApplication {
```

（5）注入 UserMapper，获取数据。在 HelloWorldController 中注入 UserMapper，在 hello6 方法中使用 userMapper 获取数据库中的数据。

```
@Autowired
private UserMapper userMapper;

@RequestMapping(value = "/hello6.do",method = RequestMethod.GET)
public String hello6(Model model) {
    model.addAttribute("users", userMapper.getAll());
    return "hello";
}
```

（6）在浏览器中输入 url“http://localhost:8081/helloworld/hello6.do”，可以将数据库中的数据查询出来显示到页面中，如图 9-5 所示。

图 9-5

2. 基于 XML 引入 MyBatis

一般简单的增、删、改、查 SQL 可以使用注解来实现，但对于复杂的 SQL 还是需要使用 XML 来进行配置，这里在上面 demo 的基础上来介绍 Spring Boot 基于 XML 引入 MyBatis。如果使用 XML 进行配置，在 interface　UserMapper 中的这些注解就不能再用了，可以先把里面的注解注释。注解注释之后，需要怎么配置将 SQL 与 interface　UserMapper 中的方法对照上呢？这里在 resources 包下创建 mybatis 文件夹，在 mybatis 文件夹下创建 mybatis-config.xml 配置文件和用于映射 xml 存放 mapper 文件夹，在 mapper 文件夹中创建 UserMapper.xml（用于与 SQL 映射），interface　UserMapper 中的方法通过 UserMapper.xml 元素中的 id 相关联。

mybatis-config.xml：

```
<?xml version="1.0" encoding="UTF-8" ?>
<!DOCTYPE configuration PUBLIC "-//mybatis.org//DTD Config 3.0//EN"
"http://mybatis.org/dtd/mybatis-3-config.dtd">
<configuration>
    <typeAliases>
        <typeAlias alias="Integer" type="java.lang.Integer" />
        <typeAlias alias="Long" type="java.lang.Long" />
        <typeAlias alias="HashMap" type="java.util.HashMap" />
        <typeAlias alias="LinkedHashMap" type="java.util.LinkedHashMap" />
        <typeAlias alias="ArrayList" type="java.util.ArrayList" />
        <typeAlias alias="LinkedList" type="java.util.LinkedList" />
    </typeAliases>
</configuration>
```

UserMapper.xml：

```
<?xml version="1.0" encoding="UTF-8" ?>
<!DOCTYPE mapper PUBLIC "-//mybatis.org//DTD Mapper 3.0//EN"
"http://mybatis.org/dtd/mybatis-3-mapper.dtd" >
<mapper namespace="com.example.mapper.UserMapper" >
    <resultMap id="BaseResultMap" type="com.example.demo.User" >
        <id column="id" property="id" jdbcType="NUMERIC" />
        <result column="name" property="name" jdbcType="VARCHAR" />
        <result column="age" property="age" jdbcType="NUMERIC" />
    </resultMap>

    <sql id="Base_Column_List" >
        id, name, age
    </sql>
```

```
    <select id="getAll" resultMap="BaseResultMap"  >
        SELECT
        <include refid="Base_Column_List" />
        FROM user
    </select>

    <select id="getOne" parameterType="java.lang.Integer"
resultMap="BaseResultMap" >
         SELECT
        <include refid="Base_Column_List" />
        FROM user  WHERE id = #{id}
    </select>

    <insert id="insert" parameterType="com.example.demo.User" >
        INSERT INTO user (name,age) VALUES (#{name}, #{age})
    </insert>

    <update id="update" parameterType="com.example.demo.User" >
        UPDATE  user SET
            <if test="name != null">name = #{name},</if>
            <if test="age != null">age = #{age}</if>
        WHERE  id = #{id}
    </update>

    <delete id="delete" parameterType="java.lang.Integer" >
        DELETE FROM user  WHERE  id =#{id}
    </delete>
</mapper>
```

这些新增之后需要把新建的 xml 配置到 mybatis 中，让它起作用，所以需要在 application.properties 配置 mybatis 的一些属性，设置配置文件和映射文件的位置。

```
mybatis.type-aliases-package=com.example.demo
mybatis.config-location=classpath:mybatis/mybatis-config.xml
mybatis.mapper-locations=classpath:mybatis/mapper/*.xml
```

修改 hello6 方法，在方法中将 id=1 的 user 的 age 更改为 27 之后再查询 user。

```
@RequestMapping(value = "/hello6.do",method = RequestMethod.GET)
public String hello6(Model model) {
    User user=new User();
    user.setAge(27);
    user.setId(1);
    userMapper.update(user);
    model.addAttribute("users", userMapper.getAll());
        return "hello";
}
```

启动项目，在浏览器中输入 url“http://localhost:8081/helloworld/hello6.do”，还是可以显示出用户列表，用户 id=1 的 age 已经修改为了 27，如图 9-6 所示。

图 9-6

图 9-7 所示是项目的目录结构，方便读者在学习的时候参考。

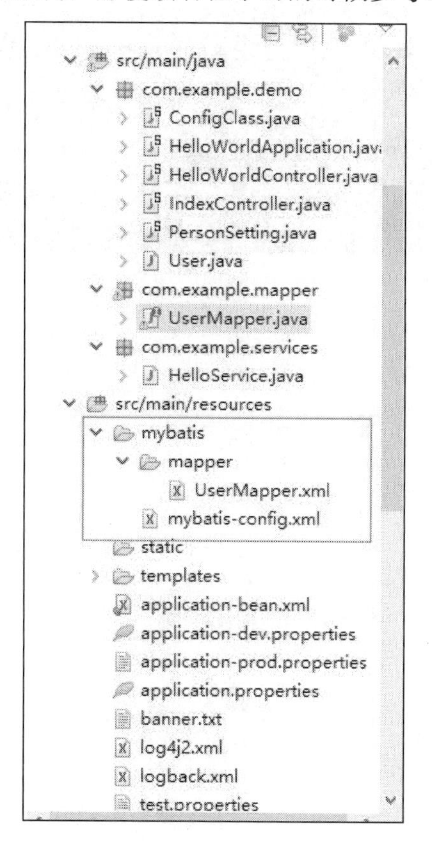

图 9-7

9.2.2　Spring Boot 集成 Redis

Redis 是一个开源的使用 ANSI C 语言编写、支持网络、可基于内存亦可持久化的日志型、Key-Value 数据库，并提供多种语言的 API。至于 Redis 的作用场景，这些暂不列举。

Lettuce 和 Jedis 的都是连接 Redis Server 的客户端程序。Jedis 在实现上是直连 Redis Server，多线程环境下非线程安全，除非使用连接池，为每个 Jedis 实例增加物理连接。Lettuce 基于 Netty 的连接实例（StatefulRedisConnection），可以在多个线程间并发访问，且线程安全，满足多线程环境下的并发访问，同时它是可伸缩的设计，一个连接实例不够的情况也可以按需增加连接实例。本节主要介绍 Spring Boot 如何使用 Lettuce 集成 Redis。为了演示集成 Redis，这里新建了一个名为 SpringBootRedis 的项目，并使用命令行 redis-server.exe redis.windows.conf 在本地启动 Redis 服务，如图 9-8 所示。

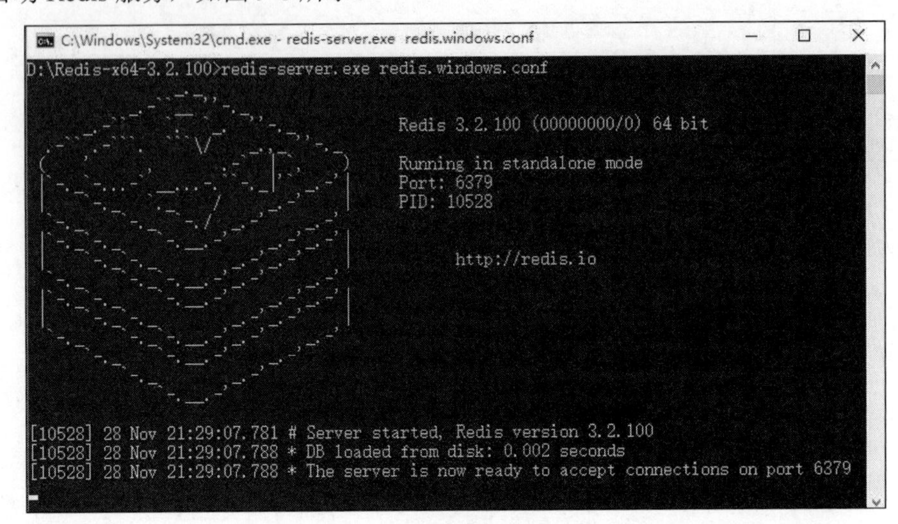

图 9-8

1. 引入依赖

访问 Redis 需要引入 spring-boot-starter-data-redis、commons-pool2 两个依赖，同时引入了集成 JSP 页面相关的依赖。

```
<dependency>
      <groupId>org.springframework.boot</groupId>
      <artifactId>spring-boot-starter-data-redis</artifactId>
</dependency>
<dependency>
      <groupId>org.apache.commons</groupId>
      <artifactId>commons-pool2</artifactId>
</dependency>
<dependency>
      <groupId>org.apache.tomcat.embed</groupId>
      <artifactId>tomcat-embed-jasper</artifactId>
      <scope>provided</scope>
</dependency>
<dependency>
      <groupId>javax.servlet</groupId>
      <artifactId>jstl</artifactId>
      <scope>provided</scope>
</dependency>
<dependency>
      <groupId>javax.servlet</groupId>
      <artifactId>javax.servlet-api</artifactId>
```

```
        <scope>provided</scope>
</dependency>
```

2. 属性配置

访问 Redis 同时还需要配置 Redis 地址、端口等信息，同时设置 Spring Boot 集成 JSP 页面的配置信息。

```
spring.redis.host=127.0.0.1
spring.redis.password=
spring.redis.port= 6379
spring.redis.timeout=1000
spring.redis.database=0
spring.redis.lettuce.pool.min-idle=0
spring.redis.lettuce.pool.max-idle=8
spring.redis.lettuce.pool.max-wait=-1ms
spring.redis.lettuce.pool.max-active=8
spring.mvc.view.prefix=/view/
spring.mvc.view.suffix=.jsp
```

3. 实例

Redis 一般都是通过 key-value 存储数据，key 一般都是 String，但是 value 可能不一样，一般有两种：String 和 Object；如果 k-v 都是 String 类型，我们可以直接用 StringRedisTemplate，这个也是官方建议的，也是最方便的，直接导入即用，无须多余配置！如果 k-v 是 Object 类型，则需要自定义 RedisTemplate。

首先在 com.example.config 包中增加 RedisCacheAutoConfiguration 类，在类中自定义 RedisTemplate。

```
@Configuration
@AutoConfigureAfter(RedisAutoConfiguration.class)
public class RedisCacheAutoConfiguration {
    @Bean
    public RedisTemplate<String, Serializable>
        redisCacheTemplate(LettuceConnectionFactory
        redisConnectionFactory) {
        RedisTemplate<String, Serializable> template = new RedisTemplate<>();
        template.setKeySerializer(new StringRedisSerializer());
        template.setValueSerializer(new
GenericJackson2JsonRedisSerializer());
        template.setConnectionFactory(redisConnectionFactory);
        return template;
    }
}
```

其次还是定义了 User 对象，同时重写了 toString()方法。

```
package com.example.model;

import java.io.Serializable;

public class User implements Serializable{

    private static final long serialVersionUID = 1L;

    private int id;

    private String name;

    private int age;

    public int getId() {
        return id;
    }
    public void setId(int id) {
        this.id = id;
    }
    public int getAge() {
        return age;
    }
    public void setAge(int age) {
        this.age = age;
    }
    public String getName() {
        return name;
    }
    public void setName(String name) {
        this.name = name;
    }


    public User(int id, String name, int age) {
        super();
        this.id = id;
        this.name = name;
        this.age = age;
    }

    public User() {
        super();
    }

    @Override
    public String toString() {
```

```
        return "User [id=" + id + ", name=" + name + ", age=" + age + "]";
    }
}
```

然后新建 IndexController，并注入 StringRedisTemplate、RedisTemplate，同时实例了 User 对象，使用 StringRedisTemplate 通过 set 方法向 Redis 中写入 User 对象的 toString()返回的字符串，并通过 get 方法获取到对应的值，使用 RedisTemplate，也是一样，只是保存的是 User 对象。

```
package com.example.demo;
import java.io.Serializable;
import org.springframework.beans.factory.annotation.Autowired;
import org.springframework.data.redis.core.RedisTemplate;
import org.springframework.data.redis.core.StringRedisTemplate;
import org.springframework.stereotype.Controller;
import org.springframework.ui.Model;
import org.springframework.web.bind.annotation.RequestMapping;
import org.springframework.web.bind.annotation.RequestMethod;

import com.example.model.User;

@Controller
@RequestMapping("/index")
public class IndexController {
    @Autowired
    private StringRedisTemplate stringRedisTemplate;

    @Autowired
    private RedisTemplate<String, Serializable> redisCacheTemplate;

    @RequestMapping(value = "/redistest.do",method = RequestMethod.GET)
    public String getallusers(Model model) {
        User user=new User(1,"小明",27);
        String strkey="stringuser";
        stringRedisTemplate.opsForValue().set(strkey, user.toString());
        final String valueStr =
stringRedisTemplate.opsForValue().get(strkey);
        model.addAttribute(strkey, valueStr);
        String objkey = "objuser";
        redisCacheTemplate.opsForValue().set(objkey, user);
        user = (User) redisCacheTemplate.opsForValue().get(objkey);
        model.addAttribute(objkey, user);
        return "index";
    }
}
```

和集成 MyBatis 时一样，在 src/main/webapp 下新建 view 文件夹，并在文件夹里面创建了 index.jsp 页面。在 index.jsp 中显示从 Redis 获取到的字符串${stringuser}和${objuser}对象。

```
<%@ page language="java" contentType="text/html; charset=utf-8"
    pageEncoding="utf-8"%>
<%@ taglib prefix="c" uri="http://java.sun.com/jsp/jstl/core"%>
<!DOCTYPE html PUBLIC "-//W3C//DTD HTML 4.01 Transitional//EN"
"http://www.w3.org/TR/html4/loose.dtd">
<html>
<head>
<meta http-equiv="Content-Type" content="text/html; charset=utf-8">
<title>Insert title here</title>
</head>
<body>
${stringuser}
  <table border="1">
        <thead>
            <tr>
               <th>序号</th>
                <th>姓名</th>
                <th>年龄</th>
            </tr>
        </thead>
            <tr>
                <td>${objuser.id}</td>
                <td>${objuser.name}</td>
                <td>${objuser.age}</td>
            </tr>
  </table>
 </body>
 </html>
```

最后在浏览器中输入 url“http://localhost:8080/index/redistest.do”，可以把从 Redis 中获取到的字符串和 User 对象显示到页面中，如图 9-9 所示。

图 9-9

9.2.3 Spring Boot 集成 MyBatis 使用 Redis 做缓存

上一小节介绍了 Spring Boot 集成 Redis，本小节学习一下 MyBatis 操作中使用 Redis 做缓存。这里其实主要学习几个注解：@CachePut、@Cacheable、@CacheEvict、@CacheConfig。

1. @Cacheable

@Cacheable 的作用主要针对方法配置，能够根据方法的请求参数对其结果进行缓存，参数如表 9-1 所示。

表 9-1 @Cacheable 的参数

参　数	解　释	例　子
value	缓存的名称，在 Spring 配置文件中定义，必须指定至少一个	@Cacheable(value="mycache") @Cacheable(value={"cache1", "cache2"}
key	缓存的 key，可以为空。如果指定，就要按照 SpEL 表达式编写；如果不指定，那么默认按照方法的所有参数进行组合	@Cacheable(value="testcache", key="#userName")
condition	缓存的条件，可以为空，使用 SpEL 编写，返回 true 或者 false，只有为 true 才进行缓存	@Cacheable(value="testcache", condition="#userName.length()>2")

2. @CachePut

@CachePut 的作用主要针对方法配置，能够根据方法的返回值对其结果进行缓存。和 @Cacheable 不同的是，它每次都会触发真实方法的调用，其他地方写的是根据方法的请求参数对其结果进行缓存，实际是对方法返回值进行缓存的，参数如表 9-2 所示。

表 9-2 @CachePut 的参数

参　数	解　释	例　子
value	缓存的名称，在 Spring 配置文件中定义，必须指定至少一个	@CachePut(value="my cache")
key	缓存的 key，可以为空。如果指定，就要按照 SpEL 表达式编写，如果不指定，那么默认按照方法的所有参数进行组合	@CachePut(value="testcache", key="#userName")
condition	缓存的条件，可以为空，使用 SpEL 编写，返回 true 或者 false，只有为 true 才进行缓存	@CachePut(value="testcache", condition="#userName.length()>2")

3. @CachEvict

@CachEvict 的作用主要针对方法配置，能够根据一定的条件对缓存进行清空，参数如表 9-3 所示。

表 9-3 @CachEvict 的参数

参　数	解　释	例　子
value	缓存的名称，在 Spring 配置文件中定义，必须指定至少一个	@CacheEvict(value="my cache")
key	缓存的 key，可以为空。如果指定，就要按照 SpEL 表达式编写；如果不指定，那么默认按照方法的所有参数进行组合	@CacheEvict(value="testcache", key="#userName")

（续表）

参　数	解　释	例　子
condition	缓存的条件，可以为空，使用 SpEL 编写，返回 true 或者 false，只有为 true 才进行缓存	@CacheEvict(value="testcache", condition="#userName.length()>2")
allEntries	是否清空所有缓存内容，默认为 false，如果指定为 true，则方法调用后将立即清空所有缓存	@CachEvict(value="testcache", allEntries=true)
beforeInvocation	是否在方法执行前就清空，默认为 false，如果指定为 true，则在方法还没有执行的时候就清空缓存，默认情况下，如果方法执行抛出异常，则不会清空缓存	@CachEvict(value="testcache", beforeInvocation=true)

4. @CacheConfig

所有的@Cacheable()里面都有一个 value="xxx"的属性。显然，如果方法多了，写起来也是挺累的，如果可以一次性声明完就省事了。有了@CacheConfig 这个配置（@CacheConfig 是类级别的注解，允许通过@CacheConfig(cacheNames="user")这种方式设置缓存的名字），如果在你的方法中写别的名字，那么依然以方法的名字为准。

下面用一个实例来进一步了解几个注解的使用。实例是在上一节 Spring Boot 集成 Redis 的基础上进行修改的。

首先，引入 MyBatis 和 MySQL 的相关依赖，并在 application.properties 中设置 MyBatis 相关的属性。

```
<dependency>
    <groupId>org.mybatis.spring.boot</groupId>
    <artifactId>mybatis-spring-boot-starter</artifactId>
    <version>1.3.2</version>
</dependency>
<dependency>
    <groupId>mysql</groupId>
    <artifactId>mysql-connector-java</artifactId>
    <version>8.0.11</version>
</dependency>
```

```
spring.datasource.driverClassName = com.mysql.cj.jdbc.Driver
spring.datasource.url=jdbc:mysql://127.0.0.1:3306/mybatis?useUnicode=tru
e&characterEncoding=UTF-8&serverTimezone=UTC&useSSL=true
spring.datasource.username = root
spring.datasource.password = 123456
```

然后，新增 UserMapper 将 Model 与数据库 SQL 关联，在 MyBatis 章节已有提过。

```
package com.example.mapper;

import java.util.List;
import org.apache.ibatis.annotations.Delete;
import org.apache.ibatis.annotations.Insert;
```

```
import org.apache.ibatis.annotations.Mapper;
import org.apache.ibatis.annotations.Result;
import org.apache.ibatis.annotations.ResultMap;
import org.apache.ibatis.annotations.Results;
import org.apache.ibatis.annotations.Select;
import org.apache.ibatis.annotations.Update;
import org.apache.ibatis.type.JdbcType;
import com.example.model.User;

@Mapper
public interface UserMapper{

   @Select("SELECT * FROM user")
   @Results(id="userMap",value={

       @Result(property = "id",column = "id",jdbcType=JdbcType.INTEGER),

       @Result(property = "age",column = "age",jdbcType=JdbcType.INTEGER),

       @Result(property = "name",column = "name",jdbcType=JdbcType.VARCHAR)

   })
   List<User> getAll();

   @Select("SELECT * FROM user WHERE id = #{id}")
   @ResultMap(value="userMap")
   User getOne(int id);

   @Insert("INSERT INTO user(name,age) VALUES(#{name}, #{age})")
   void insert(User user);

   @Update("UPDATE user SET name=#{name},age=#{age} WHERE id =#{id}")
   void update(User user);

   @Delete("DELETE FROM user WHERE id =#{id}")
   void delete(int id);
}
```

原本计划是在 Mapper 层上增加 cache 注解，但由于 update 返回值为 void，所以这里又增加了 service 层，mapper 层算是 DAO 层。使用了@MapperScan 注解来设置 mapper 接口对应的包名。使用@CacheConfig 注解设置缓存的名字。

```
package com.example.service;
import java.util.List;

import org.mybatis.spring.annotation.MapperScan;
import org.springframework.beans.factory.annotation.Autowired;
```

```
import org.springframework.cache.annotation.CacheConfig;
import org.springframework.cache.annotation.CacheEvict;
import org.springframework.cache.annotation.CachePut;
import org.springframework.cache.annotation.Cacheable;
import org.springframework.stereotype.Service;

import com.example.mapper.UserMapper;
import com.example.model.User;

@MapperScan(basePackages = "com.example.mapper")
@Service
@CacheConfig(cacheNames="user")
public class UserService {

    @Autowired
    private UserMapper userMapper;

    public List<User> getAll()
    {
        return userMapper.getAll();
    }

    @Cacheable(key = "#p0")
    public User getOne(int id)
    {
        return userMapper.getOne(id);
    }

    public void insert(User user)
    {
        userMapper.insert(user);
    }

    @CachePut(value="user",key = "#p0.id")
    public User update(User user)
    {
        userMapper.update(user);
        return user;
    }

    @CacheEvict(value="user",key ="#p0",allEntries=true)
    public void delete(int id)
    {
        userMapper.delete(id);
    }
}
```

上面创建完之后就可以在 IndexController 中注入 UserService 使用缓存注解，在 Controller 中增加了下面的几个方法，用来对 user 进行增、删、改、查。

```
@Autowired
private UserService userService;

@RequestMapping(value = "/alluser.do",method = RequestMethod.GET)
public String getall(Model model) {
    List<User> users=userService.getAll();
    model.addAttribute("users", users);
    return "index";
}
@RequestMapping(value = "/insert.do",method = RequestMethod.GET)
public String adduser(Model model) {
    User user=new User();
    user.setName("cuiyw");
    user.setAge(27);
    userService.insert(user);
    return "forward:/index/alluser.do";
}
@RequestMapping(value = "/getuserbyid.do/{id}",method = RequestMethod.GET)
public ModelAndView GetUserById(@PathVariable("id") int id) {
    System.out.println(id);
    User user=userService.getOne(id);
    System.out.println(user.toString());
    ModelAndView modelAndView = new ModelAndView("index");
    List<User> users=new ArrayList<User>();
    users.add(user);
    modelAndView.addObject("users", users);
    return modelAndView;
}
@RequestMapping(value = "/deleteuserbyid.do/{id}",method =
RequestMethod.GET)
public String DeleteUserById(@PathVariable("id") int id) {
    userService.delete(id);
    return "forward:/index/alluser.do";
}
@RequestMapping(value = "/updateuserbyid.do/{id}",method =
RequestMethod.GET)
public String UpdateUserByid(@PathVariable("id") int id) {
    User user=userService.getOne(id);
    System.out.println(user.toString());
    user.setAge(28);
    System.out.println(user.toString());
    userService.update(user);
    System.out.println(user.toString());
```

```
        return "forward:/index/alluser.do";
    }
```

最后也是经常遗漏的一步，设置@EnableCaching。需要在 main 方法上面设置启动缓存。

```
    @ComponentScan(basePackages={"com.example.config","com.example.demo","co
m.example.service"})
    @SpringBootApplication
    @EnableCaching
    public class SpringBootRedisApplication {

       public static void main(String[] args) {
           SpringApplication.run(SpringBootRedisApplication.class, args);
       }
    }
```

启动程序，在浏览器输入 Url“http://localhost:8080/index/getuserbyid.do/1”，之后启动 Redis 客户端，输入命令 keys *来查看 Redis 缓存中的数据，如图 9-10、图 9-11 所示。

图 9-10

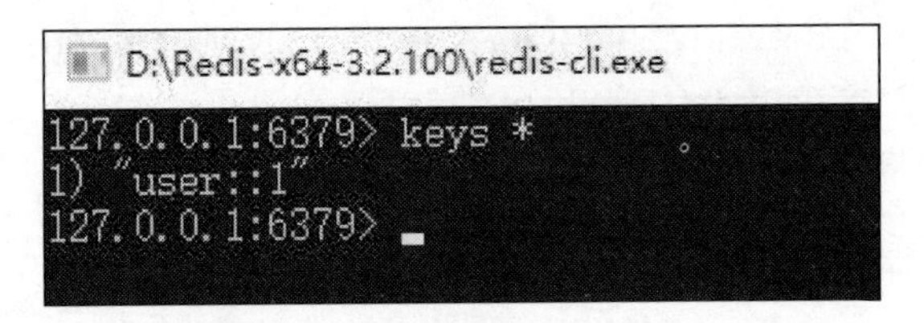

图 9-11

9.2.4 Spring Boot 集成 RabbitMQ

1. RabbitMQ 的介绍

RabbitMQ 是消息中间件的一种。消息中间件即分布式系统中完成消息的发送和接收的基础软件。消息中间件的工作过程可以用生产者消费者模型来表示，即生产者不断地向消息队列发送信息，而消费者从消息队列中消费信息，具体过程如图 9-12 所示。

从图 9-12 可看出，对于消息队列来说，生产者、消息队列、消费者是最重要的三个概念，生产者发消息到消息队列中去，消费者监听指定的消息队列，并且当消息队列收到消息之后，接收消息队列传来的消息，并且给予相应的处理。消息队列常用于分布式系统之间互相信息的传递。

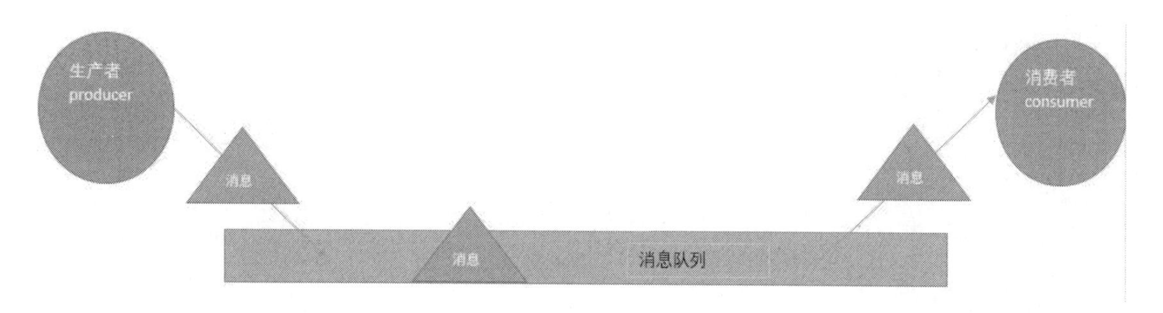

图 9-12

对于 RabbitMQ 来说，除了这三个基本模块以外，还添加了一个模块，即交换机（Exchange）。它使得生产者和消息队列之间产生了隔离，生产者将消息发送给交换机，而交换机则根据调度策略把相应的消息转发给对应的消息队列。

交换机的主要作用是接收相应的消息并且绑定到指定的队列。交换机有四种类型，分别为 Direct、topic、headers、Fanout。

Direct 是 RabbitMQ 默认的交换机模式，也是最简单的模式。在创建消息队列的时候，指定一个 BindingKey。当发送者发送消息的时候，指定对应的 Key。当 Key 和消息队列的 BindingKey 一致的时候，消息将会被发送到该消息队列中。

topic 转发信息主要是依据通配符，队列和交换机的绑定主要是依据一种模式（通配符+字符串），而当发送消息的时候，只有指定的 Key 和该模式相匹配的时候消息才会被发送到该消息队列中。

headers 也是根据一个规则进行匹配，在消息队列和交换机绑定的时候会指定一组键值对规则，而发送消息的时候也会指定一组键值对规则，当两组键值对规则相匹配的时候，消息会被发送到匹配的消息队列中。

Fanout 是路由广播的形式，将会把消息发给绑定它的全部队列，即便设置了 key，也会被忽略。

2. Spring Boot 整合 RabbitMQ(Direct 模式)

Spring Boot 整合 RabbitMQ 非常简单，首先还是 pom.xml 引入依赖。

```
<dependency>
    <groupId>org.springframework.boot</groupId>
    <artifactId>spring-boot-starter-amqp</artifactId>
</dependency>
```

在 application.properties 中配置 RabbitMQ 相关的信息，并首先启动 RabbitMQ 实例，创建两个 queue，如图 9-13 所示。

hello	D	idle	0	0	0	0.00/s	0.00/s	0.00/s
object	D	idle	0	0	0	0.00/s	0.00/s	0.00/s

图 9-13

```
spring.application.name=spirng-boot-rabbitmq
spring.rabbitmq.host=127.0.0.1
spring.rabbitmq.port=5672
spring.rabbitmq.username=admin
spring.rabbitmq.password=admin
```

配置 Queue（消息队列），由于采用的是 Direct 模式，因此需要在配置 Queue 的时候指定一个键，使其和交换机绑定。

```
@Configuration
public class RabbitConfig {

    @Bean
    public org.springframework.amqp.core.Queue Queue() {

        return new org.springframework.amqp.core.Queue("hello");

    }
}
```

接着就可以发送消息啦。在 Spring Boot 中，我们使用 AmqpTemplate 去发送消息。代码如下：

```
@Component
public class HelloSender {

    @Autowired
    private AmqpTemplate rabbitTemplate;

    public void send(int index) {

        String context = "hello Queue "+index + new Date();
        System.out.println("Sender : " + context);
        this.rabbitTemplate.convertAndSend("hello", context);
    }
}
```

生产者发送消息之后，需要消费者接收消息。这里定义了两个消息消费者，用来模拟生产者与消费者一对多的关系。

```
@Component
@RabbitListener(queues = "hello")
public class HelloReceiver {

    @RabbitHandler
    public void process(String hello) {
        System.out.println("Receiver1  : " + hello);
    }
```

```
}
@Component
@RabbitListener(queues = "hello")
public class HelloReceiver2 {

    @RabbitHandler
    public void process(String hello) {
        System.out.println("Receiver2  : " + hello);
    }
}
```

在单元测试中模拟发送消息，批量发送 10 条消息，两个接收者分别接收了 5 条消息，如图 9-14 所示。

```
@Autowired
private HelloSender helloSender;
@Test
public void hello() throws Exception {
    for(int i=0;i<10;i++)
    {
        helloSender.send(i);
    }
}
```

```
2018-12-16 17:04:10.880  INFO 14684 --- [           main] Spr
Sender : hello Queue 0Sun Dec 16 17:04:11 CST 2018
Sender : hello Queue 1Sun Dec 16 17:04:11 CST 2018
Sender : hello Queue 2Sun Dec 16 17:04:11 CST 2018
Sender : hello Queue 3Sun Dec 16 17:04:11 CST 2018
Sender : hello Queue 4Sun Dec 16 17:04:11 CST 2018
Sender : hello Queue 5Sun Dec 16 17:04:11 CST 2018
Sender : hello Queue 6Sun Dec 16 17:04:11 CST 2018
Sender : hello Queue 7Sun Dec 16 17:04:11 CST 2018
Sender : hello Queue 8Sun Dec 16 17:04:11 CST 2018
Sender : hello Queue 9Sun Dec 16 17:04:11 CST 2018
Receiver1  : hello Queue 0Sun Dec 16 17:04:11 CST 2018
Receiver2  : hello Queue 1Sun Dec 16 17:04:11 CST 2018
Receiver1  : hello Queue 2Sun Dec 16 17:04:11 CST 2018
Receiver1  : hello Queue 4Sun Dec 16 17:04:11 CST 2018
Receiver2  : hello Queue 3Sun Dec 16 17:04:11 CST 2018
Receiver1  : hello Queue 6Sun Dec 16 17:04:11 CST 2018
Receiver2  : hello Queue 5Sun Dec 16 17:04:11 CST 2018
Receiver1  : hello Queue 8Sun Dec 16 17:04:11 CST 2018
Receiver2  : hello Queue 7Sun Dec 16 17:04:11 CST 2018
Receiver2  : hello Queue 9Sun Dec 16 17:04:11 CST 2018
```

图 9-14

实际上 RabbitMQ 还可以支持发送对象，由于涉及序列化和反序列化，因此该对象要实现 Serilizable 接口。这里定义 User 对象，用来做发送消息的内容。

```
import java.io.Serializable;
public class User implements Serializable{

    private String name;
    private String pwd;
```

```
    public String getPwd() {
        return pwd;
    }
    public void setPwd(String pwd) {
        this.pwd = pwd;
    }
    public String getName() {
        return name;
    }
    public void setName(String name) {
        this.name = name;
    }
    public User(String name, String pwd) {
        this.name = name;
        this.pwd = pwd;
    }
    @Override
    public String toString() {
        return "User{" +"name='" + name + '\'' +", pwd='" + pwd + '\'' +'}';
    }
}
```

在生产者中发送 User 对象。

```
@Component
public class ModelSender {

    @Autowired
    private AmqpTemplate rabbitTemplate;

    public void sendModel(User user) {

        System.out.println("Sender object: " + user.toString());
        this.rabbitTemplate.convertAndSend("object", user);

    }
}
```

在消费者中接收 User 对象。

```
@Component
@RabbitListener(queues = "object")
public class ModelRecevicer {

    @RabbitHandler
    public void process(User user) {
```

```
        System.out.println("Receiver object : " + user);

    }
}
```

在单元测试中注入 ModelSender 对象，实例化 User 对象，然后发送，如图 9-15 所示。

```
@Autowired
private ModelSender modelSender;
@Test
public void model() throws Exception {

    User user=new User("abc","123");
    modelSender.sendModel(user);
}
```

```
Sender object: User{name='abc', pwd='123'}
2018-12-16 17:00:54.174  INFO 12984 --- [          Thr
2018-12-16 17:00:54.211  INFO 12984 --- [          Thr
Receiver object : User{name='abc', pwd='123'}
```

图 9-15

3. Spring Boot 整合 RabbitMQ(Topic 转发模式)

首先需要在 RabbitMQ 服务端创建交换机 topicExchange，并绑定两个 queue：topic.message、topic.messages，如图 9-16 所示。

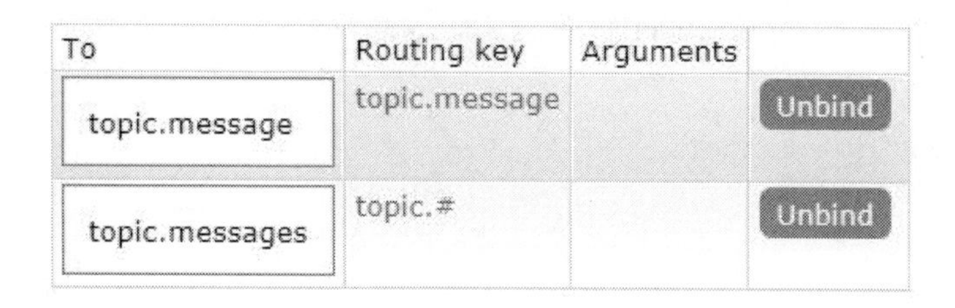

To	Routing key	Arguments	
topic.message	topic.message		Unbind
topic.messages	topic.#		Unbind

图 9-16

新建 TopicRabbitConfig，设置对应的 queue 与 binding。

```
@Configuration
public class TopicRabbitConfig {

    final static String message = "topic.message";
    final static String messages = "topic.messages";

    @Bean
    public Queue queueMessage() {
        return new Queue(TopicRabbitConfig.message);
    }
```

```
        @Bean
        public Queue queueMessages() {
            return new Queue(TopicRabbitConfig.messages);
        }

        @Bean
        TopicExchange exchange() {
            return new TopicExchange("topicExchange");
        }

        @Bean
        Binding bindingExchangeMessage(Queue queueMessage, TopicExchange
exchange) {
            return
BindingBuilder.bind(queueMessage).to(exchange).with("topic.message");
        }

        @Bean
        Binding bindingExchangeMessages(Queue queueMessages, TopicExchange
exchange) {
            return BindingBuilder.bind(queueMessages).to(exchange).
with("topic.#");
        }
    }
```

创建消息生产者，在 TopicSender 中发送 3 个消息。

```
    @Component
    public class TopicSender {

        @Autowired
        private AmqpTemplate rabbitTemplate;

        public void send() {
            String context = "hi, i am message all";
            System.out.println("Sender : " + context);
            this.rabbitTemplate.convertAndSend("topicExchange", "topic.1",
context);
        }

        public void send1() {
            String context = "hi, i am message 1";
            System.out.println("Sender : " + context);
            this.rabbitTemplate.convertAndSend("topicExchange",
"topic.message", context);
        }
```

```
    public void send2() {
        String context = "hi, i am messages 2";
        System.out.println("Sender : " + context);
        this.rabbitTemplate.convertAndSend("topicExchange",
"topic.messages", context);
    }
}
```

生产者发送消息，这里创建了两个接收消息的消费者。

```
@Component
@RabbitListener(queues = "topic.message")
public class TopicReceiver {

    @RabbitHandler
    public void process(String message) {
        System.out.println("Topic Receiver1  : " + message);
    }
}

@Component
@RabbitListener(queues = "topic.messages")
public class TopicReceiver2 {

    @RabbitHandler
    public void process(String message) {

        System.out.println("Topic Receiver2  : " + message);

    }
}
```

在单元测试中注入 TopicSender，利用 topicSender 发送消息，如图 9-17 所示。

```
@Autowired
private TopicSender topicSender;
@Test
public void topicSender() throws Exception {
    topicSender.send();
    topicSender.send1();
    topicSender.send2();
}
```

```
Sender : hi, i am message all
Sender : hi, i am message 1
Sender : hi, i am messages 2
2018-12-16 18:57:17.284  INFO 11412 --- [
Topic Receiver2  : hi, i am message all
Topic Receiver1  : hi, i am message 1
Topic Receiver2  : hi, i am message 1
Topic Receiver2  : hi, i am messages 2
```

图 9-17

从上面的输出结果可以看到，Topic Receiver2 匹配到了所有消息，Topic Receiver1 只匹配到了 1 个消息。

4. Spring Boot 整合 RabbitMQ（Fanout Exchange 形式）

Fanout Exchange 形式又叫广播形式，因此我们发送到路由器的消息会使得绑定到该路由器的每一个 Queue 接收到消息。首先需要在 RabbitMQ 服务端创建交换机 fanoutExchange，并绑定三个 queue：fanout.A、fanout.B、fanout.C，如图 9-18 所示。

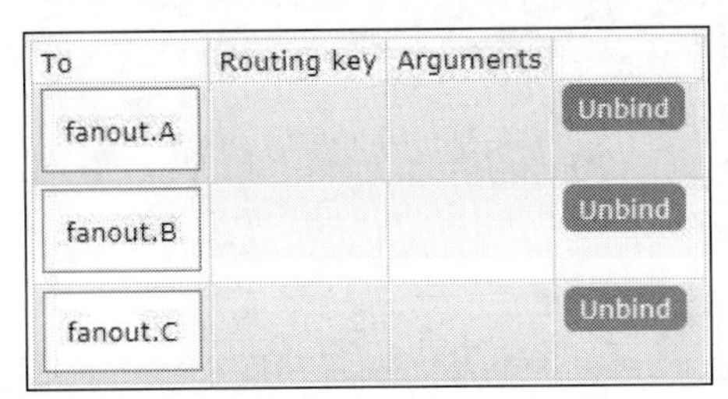

To	Routing key	Arguments	
fanout.A			Unbind
fanout.B			Unbind
fanout.C			Unbind

图 9-18

与 Topic 类似，新建 FanoutRabbitConfig，绑定交换机和队列。

```
@Configuration
public class FanoutRabbitConfig {

    @Bean
    public Queue AMessage() {
        return new Queue("fanout.A");
    }

    @Bean
    public Queue BMessage() {
        return new Queue("fanout.B");
    }

    @Bean
    public Queue CMessage() {
        return new Queue("fanout.C");
    }

    @Bean
    FanoutExchange fanoutExchange() {
        return new FanoutExchange("fanoutExchange");
    }

    @Bean
    Binding bindingExchangeA(Queue AMessage,FanoutExchange fanoutExchange){
        return BindingBuilder.bind(AMessage).to(fanoutExchange);
    }

    @Bean
    Binding bindingExchangeB(Queue BMessage, FanoutExchange
fanoutExchange){
```

```
        return BindingBuilder.bind(BMessage).to(fanoutExchange);
    }

    @Bean
    Binding bindingExchangeC(Queue CMessage, FanoutExchange
fanoutExchange){
        return BindingBuilder.bind(CMessage).to(fanoutExchange);
    }
}
```

创建消息生产者，在 FanoutSender 中发送消息。

```
@Component
public class FanoutSender {

    @Autowired
    private AmqpTemplate rabbitTemplate;

    public void send() {

        String context = "hi, fanout msg ";
        System.out.println("FanoutSender : " + context);
        this.rabbitTemplate.convertAndSend("fanoutExchange","", context);

    }
}
```

然后创建 3 个接收者：FanoutReceiverA、FanoutReceiverB、FanoutReceiverC。

```
@Component
@RabbitListener(queues = "fanout.A")
public class FanoutReceiverA {

    @RabbitHandler
    public void process(String message) {
        System.out.println("fanout Receiver A  : " + message);
    }
}
@Component
@RabbitListener(queues = "fanout.B")
public class FanoutReceiverB {

    @RabbitHandler
    public void process(String message) {
        System.out.println("fanout Receiver B: " + message);
    }
}
@Component
@RabbitListener(queues = "fanout.C")
public class FanoutReceiverC {

    @RabbitHandler
```

```
    public void process(String message) {
        System.out.println("fanout Receiver C: " + message);
    }
}
```

在单元测试中注入消息发送者，发送消息。

```
@Autowired
private FanoutSender fanoutSender;
@Test
public void fanoutSender() throws Exception {
        fanoutSender.send();
}
```

从图 9-19 可以看到 3 个队列都接收到了消息。

```
FanoutSender : hi, fanout msg
2018-12-16 19:36:44.629  INFO 16324 ---
fanout Receiver C: hi, fanout msg
fanout Receiver B: hi, fanout msg
fanout Receiver A  : hi, fanout msg
```

图 9-19

本节创建的类比较多，结构如图 9-20 所示。也可以直接查看 demo 源码了解。

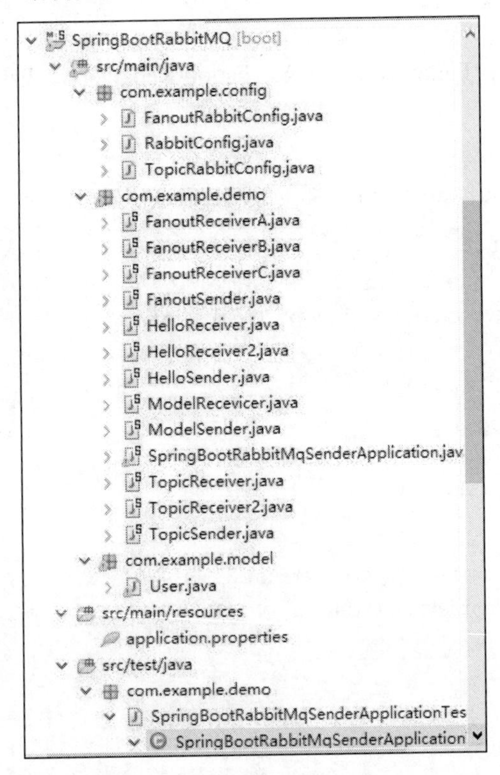

图 9-20

9.3 Spring Boot 之集成其他工具

在上一节介绍了 Spring Boot 集成 Data 层，本节介绍开发常用的几个工具，主要包括 Druid 集成、打包部署、定时任务、邮件发送。

9.3.1 Spring Boot 集成 Druid

Druid 是阿里巴巴开源平台上的一个数据库连接池实现，结合了 C3P0、DBCP、PROXOOL 等 DB 池的优点，同时加入了日志监控，可以很好地监控 DB 池连接和 SQL 的执行情况，可以说是针对监控而生的 DB 连接池。

1. 引入依赖

创建 SpringBootDruid 项目，在项目中实现 JSP 页面，显示通过 MyBatis 从 MySQL 数据库中查询出来的数据，具体实现代码就不贴上来了。为了演示 Druid 的使用，引入了 druid-spring-boot-starter。

```
<dependency>
     <groupId>com.alibaba</groupId>
     <artifactId>druid-spring-boot-starter</artifactId>
     <version>1.1.10</version>
</dependency>
```

2. 设置 Druid 属性

```
    spring.datasource.druid.type=com.alibaba.druid.pool.DruidDataSource
    spring.datasource.druid.max-active=20
    spring.datasource.druid.initial-size=1
    spring.datasource.druid.max-wait=60000
    spring.datasource.druid.pool-prepared-statements=true
    spring.datasource.druid.max-pool-prepared-statement-per-connection-size=
20
    spring.datasource.druid.connection-properties=druid.stat.mergeSql=true;d
ruid.stat.slowSqlMillis=5000
    spring.datasource.druid.min-idle=1
    spring.datasource.druid.time-between-eviction-runs-millis=60000
    spring.datasource.druid.min-evictable-idle-time-millis=300000
    spring.datasource.druid.validation-query=select 1 from dual
    spring.datasource.druid.test-while-idle=true
    spring.datasource.druid.test-on-borrow=true
    spring.datasource.druid.test-on-return=true

    spring.datasource.druid.web-stat-filter.enabled=true
    spring.datasource.druid.web-stat-filter.url-pattern=/*
```

```
spring.datasource.druid.web-stat-filter.exclusions=*.js,*.gif,*.jpg,*.png,*.css,*.ico,/druid/*
spring.datasource.druid.web-stat-filter.session-stat-enable=true
spring.datasource.druid.web-stat-filter.session-stat-max-count=1000

spring.datasource.druid.stat-view-servlet.enabled= true
spring.datasource.druid.stat-view-servlet.url-pattern=/druid/*
spring.datasource.druid.stat-view-servlet.reset-enable=true
spring.datasource.druid.stat-view-servlet.login-username=druid
spring.datasource.druid.stat-view-servlet.login-password=123456
spring.datasource.druid.stat-view-servlet.allow=127.0.0.1
spring.datasource.druid.stat-view-servlet.deny=192.168.0.19
```

3. Druid 测试

先启动应用，然后在浏览器输入“http://127.0.0.1:8080/druid/index.html”，就会显示登录页面，输入上面配置的密码即可登录，如图 9-21 所示。

图 9-21

在浏览器访问 Controller 中的方法之后就可以在 Druid 中看到 SQL 监控记录，如图 9-22 所示。

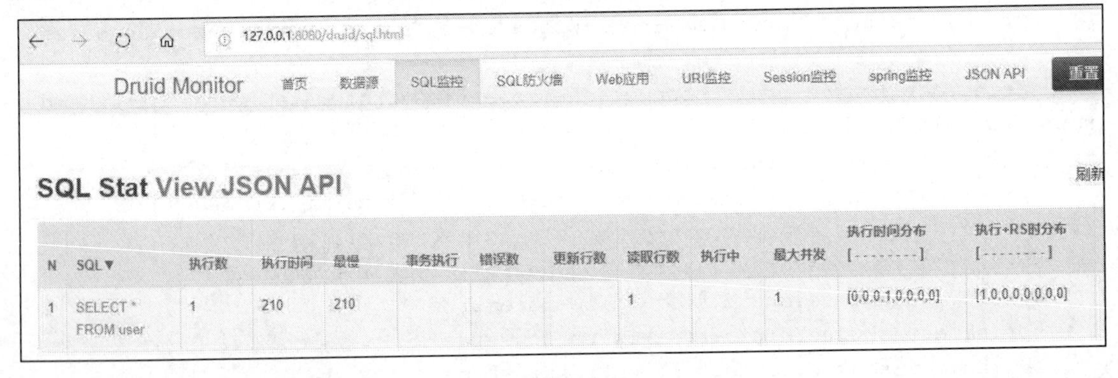

图 9-22

9.3.2 Spring Boot 定时任务

定时任务有好多开源框架，比如 Quartz。@Scheduled 是 Spring 的一个定时任务注解，通过注解配置就能够完成轻量级的定时任务，简单方便。@Scheduled 注解主要包含 cron、fixedRate、fixedDelay、initialDelay 属性。

（1）cron 属性

这是一个时间表达式，可以通过简单的配置完成各种时间的配置。我们通过 CRON 表达式几乎可以完成任意的时间搭配，它包含了 6 个或 7 个域：

- Seconds：可出现", - * /"四个字符，有效范围为 0～59 的整数。
- Minutes：可出现", - * /"四个字符，有效范围为 0～59 的整数。
- Hours：可出现", - * /"四个字符，有效范围为 0～23 的整数。
- DayofMonth：可出现", - * / ? L W C"八个字符，有效范围为 0～31 的整数。
- Month：可出现", - * /"四个字符，有效范围为 1～12 的整数或 JAN-DEc。
- DayofWeek：可出现", - * / ? L C #"四个字符，有效范围为 1～7 的整数或 SUN-SAT 两个范围。1 表示星期天，2 表示星期一，以此类推。
- Year：可出现", - * /"四个字符，有效范围为 1970～2099 年。
  - "0 0 12 * * ?"：每天中午十二点触发。
  - "0 15 10 ? * *"：每天早上 10：15 触发。
  - "0 15 10 * * ?" ：每天早上 10：15 触发。
  - "0 15 10 * * ? *"：每天早上 10：15 触发。
  - "0 15 10 * * ? 2005"：2005 年的每天早上 10：15 触发。
  - "0 * 14 * * ?"：每天从下午 2 点开始到 2 点 59 分每分钟一次触发。
  - "0 0/5 14 * * ?"：每天从下午 2 点开始到 2：55 分结束每 5 分钟一次触发。
  - "0 0/5 14,18 * * ?"：每天的下午 2 点至 2：55 和 6 点至 6 点 55 分两个时间段内每 5 分钟一次触发。
  - "0 0-5 14 * * ?"：每天 14:00 至 14:05 每分钟一次触发。
  - "0 10,44 14 ? 3 WED"：三月的每周三的 14：10 和 14：44 触发。
  - "0 15 10 ? * MON-FRI"：每个周一、周二、周三、周四、周五的 10：15 触发。

（2）fixedRate 属性

上一个调用开始后再次调用的延时（不用等待上一次调用完成），这样就会存在重复执行的问题，所以不建议使用，但数据量如果不大时在配置的间隔时间内可以执行完也是可以使用的。

（3）fixedDelay 属性

该属性的功效与上面的 fixedRate 是相反的，配置了该属性后会等到方法执行完成后延迟配置的时间再次执行该方法。

（4）initialDelay 属性

该属性跟上面的 fixedDelay、fixedRate 有着密切的关系，为什么这么说呢？该属性的作用是第一次执行延迟时间，只是做延迟的设定，并不会控制其他逻辑，所以要配合 fixedDelay 或者 fixedRate 来使用。

这里创建了 SpringBootScheduled 项目，在项目中新增了 SchedulerTask 类。在 SchedulerTask 类中定义了 3 个方法来测试上面 4 个属性的使用。

```
package com.example.demo;
import java.text.SimpleDateFormat;
import java.util.Date;
import org.springframework.scheduling.annotation.Scheduled;
import org.springframework.stereotype.Component;

@Component
public class SchedulerTask {
    SimpleDateFormat dateFormat = new SimpleDateFormat("HH:mm:ss");

    @Scheduled(fixedRate = 10000)
    public void timerRate() {
        System.out.println(dateFormat.format(new Date()));
    }

    //第一次延迟 1 秒执行，当执行完后 2 秒再执行
    @Scheduled(initialDelay = 1000, fixedDelay = 2000)
    public void timerInit() {
        System.out.println("init : "+dateFormat.format(new Date()));
    }

    //每天 21 点 41 分 50 秒执行
    @Scheduled(cron = "50 41 21 * * ?")
    public void timerCron() {
        System.out.println("current time : "+ dateFormat.format(new Date()));
    }
}
```

最后还需要在 main 方法类中设置@EnableScheduling 来开启定时任务。启动项目可以在日志中看到打印的日期，如图 9-23 所示。

```
init : 23:08:57
23:08:58
init : 23:08:59
init : 23:09:01
init : 23:09:03
init : 23:09:05
init : 23:09:07
23:09:08
init : 23:09:09
init : 23:09:11
init : 23:09:13
init : 23:09:15
init : 23:09:17
23:09:18
```

图 9-23

9.3.3　Spring Boot 集成 Swagger2

Swagger 是一组围绕 OpenAPI 规范构建的开源工具，可帮助设计、构建、记录和使用 REST API。简单说下，它的出现就是为了方便进行测试后台的 restful 形式的接口，实现动态的更新，当我们在后台的接口修改了后，Swagger 可以实现自动更新，而不需要人为地维护这个接口进行测试。

Swagger 通过注解表明该接口会生成文档，包括接口名、请求方法、参数、返回信息等，下面几个是 Swagger 常用注解。

- @Api：修饰整个类，描述 Controller 的作用。
- @ApiOperation：描述一个类的一个方法，或者说一个接口。
- @ApiParam：单个参数描述。
- @ApiModel：用对象来接收参数。
- @ApiProperty：用对象接收参数时，描述对象的一个字段。
- @ApiResponse：HTTP 响应其中一个描述。
- @ApiResponses：HTTP 响应整体描述。
- @ApiIgnore：使用该注解忽略这个 API。
- @ApiError ：发生错误返回的信息。
- @ApiParamImplicitL：一个请求参数。
- @ApiParamsImplicit 多个请求参数。

集成 Swagger 也比较简单，我们可以在之前章节 demo 的基础上做修改。首先引入 Swagger 相关依赖。

```
<dependency>
    <groupId>io.springfox</groupId>
    <artifactId>springfox-swagger2</artifactId>
    <version>2.9.2</version>
</dependency>
<dependency>
    <groupId>io.springfox</groupId>
    <artifactId>springfox-swagger-ui</artifactId>
    <version>2.9.2</version>
</dependency>
```

与启动类同级包建立 Swagger2。

```
package com.example.demo;
import org.springframework.context.annotation.Bean;
import org.springframework.context.annotation.Configuration;
import springfox.documentation.builders.ApiInfoBuilder;
import springfox.documentation.builders.PathSelectors;
import springfox.documentation.builders.RequestHandlerSelectors;
import springfox.documentation.service.ApiInfo;
import springfox.documentation.service.Contact;
```

```
import springfox.documentation.spi.DocumentationType;
import springfox.documentation.spring.web.plugins.Docket;
import springfox.documentation.swagger2.annotations.EnableSwagger2;

@Configuration
@EnableSwagger2
public class Swagger2 {
   @Bean
   public Docket createRestApi() {
       return new Docket(DocumentationType.SWAGGER_2)
              .apiInfo(apiInfo())
              .select()
              //为当前包路径
              .apis(RequestHandlerSelectors.basePackage("com.example.demo"))
              .paths(PathSelectors.any())
              .build();
   }
   //构建 api 文档的详细信息函数,注意这里的注解引用的是哪个
   private ApiInfo apiInfo() {
       return new ApiInfoBuilder()
              //页面标题
              .title("Spring Boot 测试使用 Swagger2 构建RESTful API")
              //创建人
              .contact(new Contact("社会主义接班人",
"https://www.cnblogs.com/5ishare", "991843897@qq.com"))
              //版本号
              .version("1.0")
              //描述
              .description("用户管理")
              .build();
   }
}
```

通过@Configuration 注解，表明它是一个配置类，@EnableSwagger2 开启 swagger2。

apiInfo()配置一些基本的信息。apis()指定扫描的包会生成文档。在通过 createRestApi 函数创建 Docket 的 Bean 之后，apiInfo()用来创建该 Api 的基本信息（这些基本信息会展现在文档页面中）。

select()函数返回一个 ApiSelectorBuilder 实例，用来控制哪些接口暴露给 Swagger 来展现，本例采用指定扫描的包路径来定义。Swagger 会扫描该包下所有 Controller 定义的 API，并产生文档内容（除了被@ApiIgnore 指定的请求）。

在 UserController 中设置接口信息。

```
@RestController
@RequestMapping("/user")
public class UserController {

    @Autowired
    private UserMapper userMapper;

    @ApiOperation(value ="查询所有用户",notes ="",httpMethod = "GET")
    @GetMapping("/user")
    public List<User> userList(){
     return  userMapper.getAll();
    }

    /**
     * 查询用户根据 id
     * @return USer 对象
     * @param  id
     */
    @ApiOperation(value = "查询用户",notes = "根据用户 id 查询用户",httpMethod =
"GET")
    @ApiImplicitParam(name = "id",value = "用户 id",required = true,dataType
= "String")
    @GetMapping("/user/{id}")
    public User getUserById(@PathVariable String id){
        User user  = userMapper.getOne(id);
Return user;
    }
    /**
     * 新增 User
     * @param user
     * @return success or error
     */
    @ApiOperation(value = "新增用户",notes = "",httpMethod = "POST")
    @ApiImplicitParam(name = "user",value = "用户实体",required = true,
dataType = "User")
    @PostMapping("/user")
    public String createUser(@RequestBody User user){
       int num =  userMapper.insert(user);
       if(num>0){
          return "success";
       }
        return "error";
    }
    /**
     * 更新 User
```

```
     * @param user
     * @return success or error
     */
    @ApiOperation(value = "更新用户",notes = "根据用户 id 更新用户",httpMethod =
"PUT")
    @ApiImplicitParams({
            @ApiImplicitParam(name = "id",value = "用户 id",required = true,
dataType = "String"),
            @ApiImplicitParam(name = "user",value = "用户实体,传入更改后的数据
",required = true,dataType = "User")
    })
    @PutMapping("user/{id}")
    public String updateUser(@PathVariable String id,@RequestBody User
user){
        int num =  userMapper.update(user);
        if(num>0){
            return "success";
        }
        return "error";
    }

    /**
     * 删除用户
     * @param id
     * @return success or error
     */
    @ApiOperation(value = "删除用户",notes = "",httpMethod = "DELETE")
    @ApiImplicitParam(name = "id",value = "用户 id",required = true,dataType
= "String")
    @DeleteMapping("user/{id}")
    public String deleteUser(@PathVariable String id){
        int num = userMapper.delete(id);
        if(num>0){
            return "success";
        }
        return "error";
    }
}
```

启动类设置@MapperScan。@MapperScan("com.example.mapper")最后启动项目：访问 http://localhost:8080/swagger-ui.html，如图 9-24 所示。

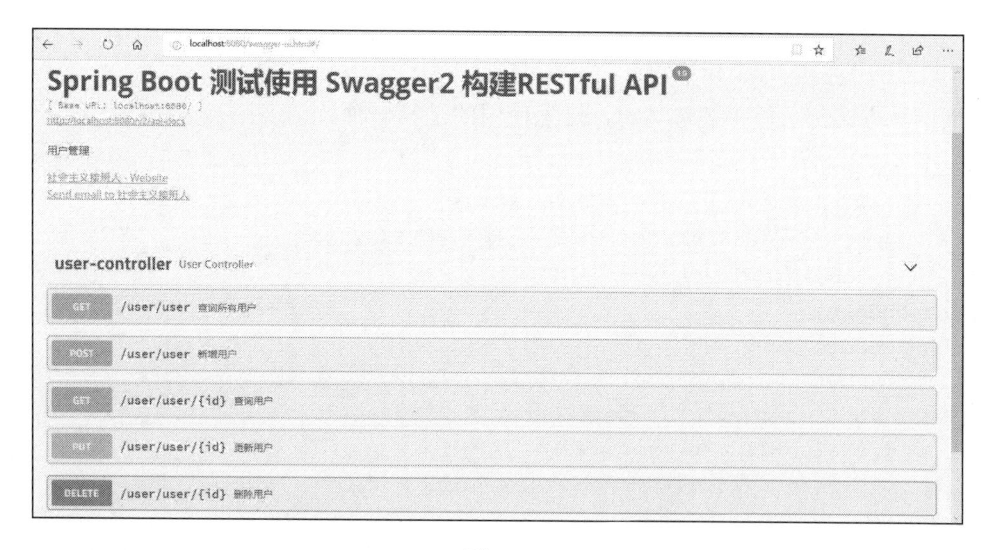

图 9-24

9.3.4　Spring Boot 打包部署

正常情况下，Spring Boot 项目是 jar 包的形式，并且内嵌 Tomcat 服务器，所以每次重新启动都是使用的新的 Tomcat 服务器。正因如此，出现了一个问题：上传到项目的文件，如果是保存在项目中的，那么重启过后文件就会丢失。比如我们上传了一个头像，重启项目后，这个头像就没了。如果将文件保存在本地磁盘中，HTML 中的标签就没有办法获取（当然，企业项目中一般是有专门的图片服务器的）。因此，我们才需要将 Spring Boot 项目打成 war 包，放到 Tomcat 中去运行。

Spring Boot 打成 war 包部署到 Tomcat 只需 5 步，这里以上一节的 SpringBootSwagger 项目为 demo。

（1）修改打包形式

在 pom.xml 里设置<packaging>war</packaging>。

（2）移除嵌入式 Tomcat 插件

在 pom.xml 里找到 spring-boot-starter-web 依赖节点，在其中添加如下代码。

```
<dependency>
    <groupId>org.springframework.boot</groupId>
    <artifactId>spring-boot-starter-web</artifactId>
    <exclusions>
        <exclusion>
            <groupId>org.springframework.boot</groupId>
            <artifactId>spring-boot-starter-tomcat</artifactId>
        </exclusion>
    </exclusions>
</dependency>
```

（3）添加 servlet-api 的依赖

下面两种方式都可以，任选其一，这里使用的是第一种。

```
<dependency>
      <groupId>javax.servlet</groupId>
      <artifactId>javax.servlet-api</artifactId>
      <scope>provided</scope>
</dependency>
<dependency>
      <groupId>org.apache.tomcat</groupId>
      <artifactId>tomcat-servlet-api</artifactId>
      <version>8.0.36</version>
      <scope>provided</scope>
</dependency>
```

（4）修改启动类，并重写初始化方法

该类继承了 SpringBootServletInitializer 并且重写了 configure 方法。

```
@SpringBootApplication
@MapperScan("com.example.mapper")
public class SpringBootSwaggerApplication extends
SpringBootServletInitializer{

   @Override
   protected SpringApplicationBuilder configure(SpringApplicationBuilder
builder) {
       return builder.sources(SpringBootSwaggerApplication.class);
   }

   public static void main(String[] args) {
       SpringApplication.run(SpringBootSwaggerApplication.class, args);
   }

}
```

（5）打包部署

在项目根目录（包含 pom.xml 的目录下），在命令行里输入“mvn clean package”即可，等待打包完成，出现[INFO] BUILD SUCCESS 即为打包成功。然后把 target 目录下的 war 包放到 tomcat 的 webapps 目录下，启动 tomcat，即可自动解压部署。最后在浏览器中输入 http://localhost:[端口号]/[打包项目名]/，如图 9-25、图 9-26、图 9-27 所示。

图 9-25

```
[INFO] --- maven-war-plugin:3.1.0:war (default-war) @ SpringBootSwagger ---
[INFO] Packaging webapp
[INFO] Assembling webapp [SpringBootSwagger] in [E:\草稿\第9章\SpringBootSwagger\target\SpringBootSw
]
[INFO] Processing war project
[INFO] Copying webapp resources [E:\草稿\第9章\SpringBootSwagger\src\main\webapp]
[INFO] Webapp assembled in [3450 msecs]
[INFO] Building war: E:\草稿\第9章\SpringBootSwagger\target\SpringBootSwagger-0.0.1-SNAPSHOT.war
[INFO]
[INFO] --- spring-boot-maven-plugin:2.0.7.RELEASE:repackage (default) @ SpringBootSwagger ---
[INFO] ------------------------------------------------------------------------
[INFO] BUILD SUCCESS
[INFO] ------------------------------------------------------------------------
[INFO] Total time: 01:45 min
[INFO] Finished at: 2019-01-06T14:30:19+08:00
[INFO] Final Memory: 38M/167M
[INFO] ------------------------------------------------------------------------
```

图 9-26

图 9-27

9.4 小结

本章主要学习 Spring Boot 的应用，介绍了 Spring Boot 与 JSP、Redis、MyBatis、Druid、RabbitMQ、Swagger2 等集成的用法，这些基本组件都是实际项目中常用到的。

第 10 章

◂ Docker入门 ▸

在上一章学习了 Spring Boot 的具体应用，在本章我们将了解一下 Docker 容器技术。

本章主要涉及的知识点：

- 镜像：镜像的创建、获取、删除等管理。
- 容器：容器的新建、启动、终止、导入导出。
- 数据管理：数据卷与数据容器的使用。
- 端口映射与容器互联：端口映射、容器互联。

10.1 Docker 基础

各大公司都在打造自己的云平台，包括阿里云、华为云、腾讯云等，以及各种微服务架构。其实，在这些当中 Docker 容器技术算是一个很重要的角色。

10.1.1 Docker 介绍

Docker 是一个开源的应用容器引擎，基于 Go 语言并遵从 Apache 2.0 协议开源。Docker 可以让开发者打包他们的应用以及依赖包到一个轻量级、可移植的容器中，然后发布到任何流行的 Linux 机器上，也可以实现虚拟化。容器完全使用沙箱机制，相互之间不会有任何接口（类似 iPhone 的 App），更重要的是容器性能开销极低。

Docker 的应用场景如下：

- Web 应用的自动化打包和发布。
- 自动化测试和持续集成、发布。
- 在服务型环境中部署和调整数据库或其他的后台应用。
- 从头编译或者扩展现有的 OpenShift 或 Cloud Foundry 平台来搭建 PaaS 环境。

Docker 优点：

（1）简化程序

Docker 让开发者可以打包他们的应用以及依赖包到一个可移植的容器中，然后发布到任何流行的 Linux 机器上，便可以实现虚拟化。Docker 改变了虚拟化的方式，使开发者可以直接将自己的成果放入 Docker 中进行管理。方便快捷已经是 Docker 的最大优势，过去需要用数天乃至数周的任务，在 Docker 容器的处理下，只需要数秒就能完成。

（2）避免选择恐惧症

如果你有选择恐惧症，还是资深患者，那么可以用 Docker 帮你打包你的纠结，比如 Docker 镜像。Docker 镜像中包含了运行环境和配置，可以简化部署多种应用实例工作，比如 Web 应用、后台应用、数据库应用、大数据应用（比如 Hadoop 集群）、消息队列等都可以打包成一个镜像部署。

（3）节省开支

一方面，云计算时代到来，使开发者不必为了追求效果而配置高额的硬件，Docker 改变了高性能必然高价格的思维定势。Docker 与云的结合，让云空间得到更充分的利用。不仅解决了硬件管理的问题，也改变了虚拟化的方式。

Docker 主要有三大核心概念：镜像、容器、仓库。

- 镜像：在安装软件操作系统的时候可能会用到镜像。Docker 中的镜像与操作系统镜像文件类似，可以理解成一个模板，有点类似手机 App 应用。
- 容器：容器是用来装东西的，Docker 中的容器用来装由镜像创建的应用运行实例。这个有点类似手机中的沙盒，每个手机 App 都用自己的运行环境，不受其他 App 的影响。
- 仓库：仓库也是用来存放东西的，是静态的，存放的是镜像文件。容器是动态的，运行的是镜像。这个有点类似 App Store，用户可以从仓库中下载 App 安装到手机中，手机中每个应用都是一个沙盒环境。

10.1.2 Docker 在 Windows 下的安装

1. 下载

在下载之前首先检查一下自己的电脑是否满足 Docker 的要求：Window 系统要是 64 位的 Windows 10 专业版，同时 Hyper-V 可用。Hyper-V 的状态设置可以是不启用状态。如果是没启用，在启动 Docker 时会有提示并会自动重启电脑来启用。下载之后就是安装，一步一步地安装。

2. 启动 Docker

在启动的时候有时会遇到如图 10-1 所示的内存不够用的提示，可以释放一些内存或是更改 Docker 设置。

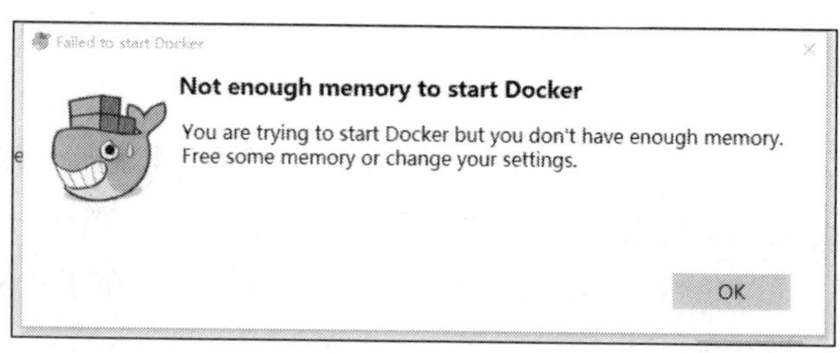

图 10-1

找到 Docker 的 setting，在设置中可以修改分配给 Docker 的资源，这里分配了 2 个 CPU、1280MB 的内存，也可以设置共享的硬盘驱动，如图 10-2 所示。

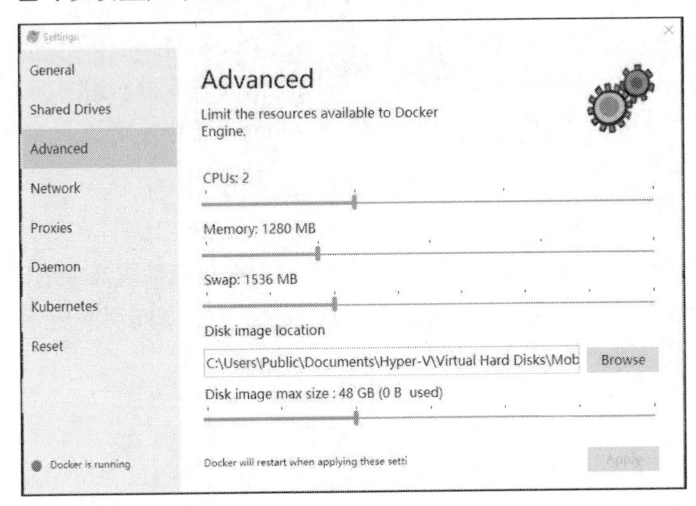

图 10-2

3. Docker 帮助文档

对于 Docker 容器算是比较新的技术，帮助文档可以直接通过命令行 docker COMMAND --help 来查看每个命令的详细介绍，如图 10-3 所示。

```
C:\WINDOWS\system32\cmd.exe
Microsoft Windows [版本 10.0.17134.523]
(c) 2018 Microsoft Corporation。保留所有权利。

C:\Users\admin>docker ps --help

Usage:  docker ps [OPTIONS]

List containers

Options:
  -a, --all             Show all containers (default shows just running)
  -f, --filter filter   Filter output based on conditions provided
      --format string   Pretty-print containers using a Go template
  -n, --last int        Show n last created containers (includes all
                        states) (default -1)
  -l, --latest          Show the latest created container (includes all
                        states)
      --no-trunc        Don't truncate output
  -q, --quiet           Only display numeric IDs
  -s, --size            Display total file sizes
```

图 10-3

4. Docker 测试

学习新语言一般都是从 hello world 开始的，Docker 也一样，可以通过 docker run hello-world 来测试拉取一个 image 从 Docker Hub 中并起一个容器 container，如图 10-4 所示。

图 10-4 也有对 docker run hello-world 命令过程的介绍，对应内容如下。

（1）Docker client 与 Docker Daemon 连接。

（2）Docker Daemon 从 Docker Hub 中拉取一个 hello-world 的镜像 image。

（3）Docker 根据镜像 image 来创建容器 container。

（4）Docker Daemon 将输入发给 Docker Client。

```
选择C:\WINDOWS\system32\cmd.exe

C:\Users\admin>docker run hello-world
Unable to find image 'hello-world:latest' locally
latest: Pulling from library/hello-world
1b930d010525: Already exists
Digest: sha256:2557e3c07ed1e38f26e389462d03ed943586f744621577a99efb77324b0fe535
Status: Downloaded newer image for hello-world:latest

Hello from Docker!
This message shows that your installation appears to be working correctly.

To generate this message, Docker took the following steps:
 1. The Docker client contacted the Docker daemon.
 2. The Docker daemon pulled the "hello-world" image from the Docker Hub.
    (amd64)
 3. The Docker daemon created a new container from that image which runs the
    executable that produces the output you are currently reading.
 4. The Docker daemon streamed that output to the Docker client, which sent it
    to your terminal.

To try something more ambitious, you can run an Ubuntu container with:
 $ docker run -it ubuntu bash
```

图 10-4

10.2 Docker 镜像

上一节对 Docker 进行了简单的介绍，包括 Docker 的三大核心概念，本节主要介绍 Docker 三大核心概念中的镜像。

10.2.1 获取镜像

Docker 运行容器前需要本地存放有对应的镜像，如果镜像没有，会从默认的仓库下载，既然是默认仓库，那肯定是公共的，当然也可以自定义自己私有的镜像仓库。使用镜像，首先得获取镜像，使用 pull 来拉取镜像文件到本地。我们可以使用 docker pull --help 了解它的格式和一些参数说明，如图 10-5 所示。

```
C:\Users\admin>docker pull --help

Usage:  docker pull [OPTIONS] NAME[:TAG|@DIGEST]

Pull an image or a repository from a registry

Options:
  -a, --all-tags                Download all tagged images in the repository
      --disable-content-trust   Skip image verification (default true)
      --platform string         Set platform if server is multi-platform
                                capable
```

图 10-5

图 10-6 所示是使用 pull 来获取 hello-world 镜像，tag 是取最新版本。

```
C:\Users\admin>docker pull hello-world:latest
latest: Pulling from library/hello-world
Digest: sha256:2557e3c07ed1e38f26e389462d03ed943586f744621577a99efb77324b0fe535
Status: Image is up to date for hello-world:latest
```

图 10-6

10.2.2 查看镜像

对于本地安装了哪些镜像，可以使用 docker images 来查看，如图 10-7 所示。

```
C:\Users\admin>docker images
REPOSITORY                  TAG          IMAGE ID       CREATED          SIZE
hello-world                 latest       fce289e99eb9   4 weeks ago      1.84kB
mongo                       latest       525bd2016729   2 months ago     383MB
rabbitmq                    management   d69a5113ceae   3 months ago     149MB
docker4w/nsenter-dockerd    latest       2f1c802f322f   3 months ago     187kB
<none>                      <none>       cae870735e91   15 months ago    187kB
```

图 10-7

关于 docker images 的具体使用，可以使用 docker images --help 命令查看，如图 10-8 所示。

```
C:\Users\admin>docker images --help

Usage:  docker images [OPTIONS] [REPOSITORY[:TAG]]

List images

Options:
  -a, --all             Show all images (default hides intermediate images)
      --digests         Show digests
  -f, --filter filter   Filter output based on conditions provided
      --format string   Pretty-print images using a Go template
      --no-trunc        Don't truncate output
  -q, --quiet           Only show numeric IDs
```

图 10-8

10.2.3 使用 tag 添加镜像标签

使用 tag 可以给镜像添加标签，类似给镜像添加一个别名。如图 10-9 所示，首先使用命令 docker tag hello-world:latest myhello-world:latest 为 hello-world:latest 镜像设置标签 myhello-world:latest。设置之后使用 docker images 查看镜像列表，则会发现 hello-world 和 myhello-world 两个镜像的 IMAGE ID 是一致的。

```
C:\Users\admin>docker tag hello-world:latest myhello-world:latest

C:\Users\admin>docker images
REPOSITORY                  TAG          IMAGE ID       CREATED          SIZE
hello-world                 latest       fce289e99eb9   4 weeks ago      1.84kB
myhello-world               latest       fce289e99eb9   4 weeks ago      1.84kB
mongo                       latest       525bd2016729   2 months ago     383MB
rabbitmq                    management   d69a5113ceae   3 months ago     149MB
docker4w/nsenter-dockerd    latest       2f1c802f322f   3 months ago     187kB
<none>                      <none>       cae870735e91   15 months ago    187kB
```

图 10-9

10.2.4 使用 inspect 查看详细信息

可以使用 inspect 命令查看镜像详细信息。对于 inspect 的详细使用，还是可以使用 help 来查看。图 10-10 所示是使用命令 docker inspect --help 来查看 inspect 参数的详细说明。

```
C:\Users\admin>docker inspect --help

Usage:  docker inspect [OPTIONS] NAME|ID [NAME|ID...]

Return low-level information on Docker objects

Options:
  -f, --format string   Format the output using the given Go template
  -s, --size            Display total file sizes if the type is container
      --type string     Return JSON for specified type
```

图 10-10

这里使用 docker inspect hello-world 来查看 hello-world 镜像的详细信息，如图 10-11 所示。

```
C:\Users\admin>docker inspect hello-world
[
    {
        "Id": "sha256:fce289e99eb9bca977dae136fbe2a82b6b7d4c372474c92
        "RepoTags": [
            "hello-world:latest",
            "myhello-world:lastest"
        ],
```

图 10-11

10.2.5　使用 history 查看镜像历史记录

使用 docker history 查看镜像的完整记录。使用 docker history hello-world 命令来查看 hello-world 的完整记录，如图 10-12 所示。

```
C:\Users\admin>docker history hello-world
IMAGE          CREATED       CREATED BY                                      SIZE     COMMENT
fce289e99eb9   4 weeks ago   /bin/sh -c #(nop)  CMD ["/hello"]               0B
<missing>      4 weeks ago   /bin/sh -c #(nop) COPY file:f77490f70ce51da2…   1.84kB
```

图 10-12

10.2.6　镜像查找

如果想查找某个镜像，可以使用 search 命令查找。对于 search 命令的使用，可以使用 help 来具体了解，如图 10-13 所示。

```
C:\Users\admin>docker search --help

Usage:  docker search [OPTIONS] TERM

Search the Docker Hub for images

Options:
  -f, --filter filter   Filter output based on conditions provided
      --format string   Pretty-print search using a Go template
      --limit int       Max number of search results (default 25)
      --no-trunc        Don't truncate output
```

图 10-13

图 10-14 所示主要查找 stars>3 且 name 包含 hello-world 的镜像。

```
C:\Users\admin>docker search -f=stars=3 hello-world
NAME                              DESCRIPTION                                       STARS     OFFICIAL   AUTOMATED
hello-world                       Hello World! (an example of minimal Dockeriz…     812       [OK]
kitematic/hello-world-nginx       A light-weight nginx container that demonstr…     120                  [OK]
tutum/hello-world                 Image to test docker deployments. Has Apache…     59                   [OK]
dockercloud/hello-world           Hello World!                                      14                   [OK]
crccheck/hello-world              Hello World web server in under 2.5 MB            6
hypriot/armhf-hello-world         Hello World! (an example of minimal Dockeriz…     6                    [OK]
marcells/aspnet-hello-world       ASP.NET vNext - Hello World                       5
armhf/hello-world                 Hello World! (an example of minimal Dockeriz…     5                    [OK]
bonomat/nodejs-hello-world        a simple nodejs hello world container             3
kornkitti/express-hello-world     Node.js Express Hello World : https://github…     3
```

图 10-14

10.2.7　删除镜像

删除镜像使用 rmi 命令。使用 rmi 时，通过标签删除镜像时，如果该镜像只有一个标签，那么在删除标签之后也会把该镜像删除；如果有多个标签，就只删除对应的标签。通过 ID 删除时，如果该镜像创建的容器存在时镜像文件无法删除，就会尝试删除所有指向该镜像的标签，然后删除该镜像文件本身，如图 10-15 所示。

```
C:\Users\admin>docker rmi --help

Usage:  docker rmi [OPTIONS] IMAGE [IMAGE...]

Remove one or more images

Options:
  -f, --force      Force removal of the image
      --no-prune   Do not delete untagged parents
```

图 10-15

1. 使用标签删除镜像

删除 myhello_world 的镜像，然后查看镜像列表，发现只有 hello-world 了，如图 10-16 所示。

```
C:\Users\admin>docker rmi myhello-world:latest
Untagged: myhello-world:latest

C:\Users\admin>docker images
REPOSITORY                 TAG          IMAGE ID       CREATED         SIZE
hello-world                latest       fce289e99eb9   4 weeks ago     1.84kB
mongo                      latest       525bd2016729   2 months ago    383MB
rabbitmq                   management   d69a5113ceae   3 months ago    149MB
docker4w/nsenter-dockerd   latest       2f1c802f322f   3 months ago    187kB
<none>                     <none>       cae870735e91   15 months ago   187kB
```

图 10-16

2. 使用镜像 ID 删除镜像

从图 10-17 可以看到，通过 ID 删除镜像 hello-world 时，报了错误，提示不能删除，因为这个镜像 fce289e99eb9 在被容器使用，需要先停掉容器 90e4296835b8。

```
C:\Users\admin>docker rmi fce289e99eb9
Error response from daemon: conflict: unable to delete fce289e99eb9 (must be forced) -
image is being used by stopped container 90e4296835b8
```

图 10-17

使用 docker ps -a 查看容器，发现会有一个容器 90e4296835b8 使用的镜像是 fce289e99eb9，如图 10-18 所示。

```
C:\Users\admin>docker ps -a
CONTAINER ID        IMAGE                     COMMAND                  CREATED             STATUS                      PORTS
                                                                                                               NAMES
90e4296835b8        hello-world               "/hello"                 2 days ago          Exited (0) 2 days ago
                                                                                                               inspiring_shirley
6768b4e7c6f5        mongo                     "docker-entrypoint.s…"   6 weeks ago         Exited (255) 2 days ago     0.0.0.0:
                                                                                                               mongo
654472d95f98        rabbitmq:management       "docker-entrypoint.s…"   7 weeks ago         Exited (255) 6 weeks ago    4369/tcp
672->5672/tcp, 0.0.0.0:15672->15672/tcp, 0.0.0.0:25672->25672/tcp, 0.0.0.0:61613->61613/tcp, 15671/tcp   rabbit
```

图 10-18

10.2.8　创建镜像

创建镜像主要有两种方法：一是基于已有镜像的容器创建，主要命令为 commit；二是基于本地模板导入，主要命令为 import。

1. 基于已有镜像的容器创建

基于已有镜像的容器创建，用到 commit 命令，首先使用 help 了解 commit 的参数信息，如图 10-19 所示。

```
C:\Users\admin>docker commit --help

Usage:  docker commit [OPTIONS] CONTAINER [REPOSITORY[:TAG]]

Create a new image from a container's changes

Options:
  -a, --author string    Author (e.g., "John Hannibal Smith
                         <hannibal@a-team.com>")
  -c, --change list      Apply Dockerfile instruction to the created image
  -m, --message string   Commit message
  -p, --pause            Pause container during commit (default true)
```

图 10-19

以 hello-world 为例，使用 docker commit -m "add a image" -a "cuiyw" 90e4296835b8 cuiyw-hello-world 命令，基于 hello-world 镜像的容器 90e4296835b8 创建了一个新的镜像 cuiyw-hello-world。使用 docker images 命令查看镜像列表时即可发现新的 cuiyw-hello-world 镜像，如图 10-20 所示。

```
C:\Users\admin>docker images
REPOSITORY                 TAG          IMAGE ID        CREATED           SIZE
cuiyw-hello-world          latest       1743012f4558    10 seconds ago    1.84kB
<none>                     <none>       002d67520092    42 minutes ago    1.85kB
hello-world                latest       fce289e99eb9    4 weeks ago       1.84kB
mongo                      latest       525bd2016729    2 months ago      383MB
rabbitmq                   management   d69a5113ceae    3 months ago      149MB
docker4w/nsenter-dockerd   latest       2f1c802f322f    3 months ago      187kB
<none>                     <none>       cae870735e91    15 months ago     187kB
```

图 10-20

2. 基于本地模板导入

基于本地模板导入用到 import 命令，首先还是使用 help 了解 import 的参数信息，如图 10-21 所示。

```
C:\Users\admin>docker import --help

Usage:  docker import [OPTIONS] file|URL|- [REPOSITORY[:TAG]]

Import the contents from a tarball to create a filesystem image

Options:
  -c, --change list       Apply Dockerfile instruction to the created image
  -m, --message string    Set commit message for imported image
```

图 10-21

既然是基于本地模板导入，就需要先有本地模板才可以，所以还需要使用 export 命令导出 tar 文件。export 命令是对容器的操作，可以使用 help 先提前了解一下，这里使用 export 命令将基于 hello-world 的镜像容器导出，之后即可在 G 盘目录下找到 hello-world.tar，此时本地模板已经存在，如图 10-22 所示。

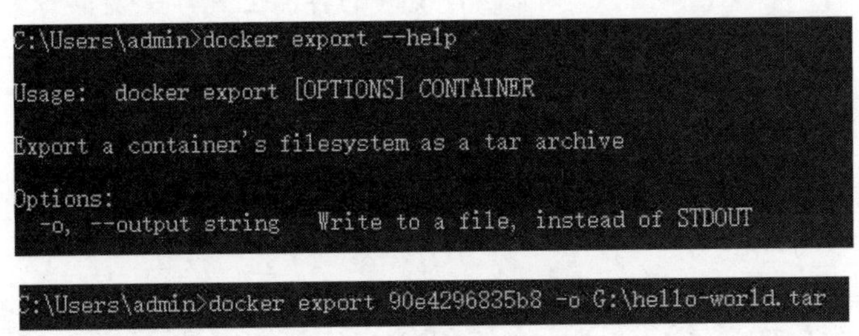

```
C:\Users\admin>docker export --help

Usage:  docker export [OPTIONS] CONTAINER

Export a container's filesystem as a tar archive

Options:
  -o, --output string   Write to a file, instead of STDOUT
```

```
C:\Users\admin>docker export 90e4296835b8 -o G:\hello-world.tar
```

图 10-22

最后使用 docker import G:\hello-world.tar hello-world-import:latest 命令将使用 export 导出的文件导入并设置新的 image 名字：hello-world-import:latest。使用 docker images 显示镜像列表时可以看到新的 image 已经存在，如图 10-23 所示。

```
C:\Users\admin>docker import G:\hello-world.tar hello-world-import:latest
sha256:6fc461936b6703d48f3dbbfcb132f5d8269968347c5377ede0ea7191f75deeb8

C:\Users\admin>docker images
REPOSITORY                  TAG          IMAGE ID       CREATED              SIZE
hello-world-import          latest       6fc461936b67   11 seconds ago       1.85kB
cuiyw-hello-world           latest       1743012f4558   19 minutes ago       1.84kB
<none>                      <none>       002d67520092   About an hour ago    1.85kB
hello-world                 latest       fce289e99eb9   4 weeks ago          1.84kB
mongo                       latest       525bd2016729   2 months ago         383MB
rabbitmq                    management   d69a5113ceae   3 months ago         149MB
docker4w/nsenter-dockerd    latest       2f1c802f322f   3 months ago         187kB
<none>                      <none>       cae870735e91   15 months ago        187kB
```

图 10-23

10.2.9 另存和载入镜像

镜像也可以以文件的形式存在，可以将镜像另存或者载入，主要用到两个命令，一个是另存命令 save，另一个是载入命令 load。我们可以先保存（save）一个镜像，然后把保存的镜像加载（load）进来。首先使用 docker save --help 来查看 save 命令的使用方法，可以从图 10-24 看到需要传两个参数，一个是-o 参数，用来指定导出的镜像文件路径，另一个是传镜像。这里还是以 hello-world 镜像为例，先使用 save 命令 docker save -o G:\helloworld-save.tar hello-world 将镜像保存到 G:\helloworld-save.tar 中。

```
C:\Users\admin>docker save --help

Usage:  docker save [OPTIONS] IMAGE [IMAGE...]

Save one or more images to a tar archive (streamed to STDOUT by default)

Options:
  -o, --output string   Write to a file, instead of STDOUT
```

```
C:\Users\admin>docker save -o G:\helloworld-save.tar hello-world
```

图 10-24

将 hello-world 镜像导出之后可以使用 load 镜像导入，还是首先使用 help 了解 load 命令的使用方法。这里使用 docker load -i G:\helloworld-save.tar 将导出的 hello-world 镜像载入进来，如图 10-25 所示。

```
Usage:  docker load [OPTIONS]

Load an image from a tar archive or STDIN

Options:
  -i, --input string   Read from tar archive file, instead of STDIN
  -q, --quiet          Suppress the load output
```

```
C:\Users\admin>docker load -i G:\helloworld-save.tar
Loaded image: hello-world:latest
```

图 10-25

10.3 容器

上一节学习了 docker 容器技术三大概念中的镜像，本节来学习 docker 容器技术的另一个概念：容器。本节主要了解容器的相关操作。

10.3.1 新建与启动容器

容器类似一个手机中的沙盒环境，用来运行 app 实例。和镜像一样也是对容器的创建、删除、导出等。创建容器可以使用 create 命令，首先还是使用 help 了解一下 create 命令的参数，如图 10-26 所示。

```
C:\Users\admin>docker create --help

Usage:  docker create [OPTIONS] IMAGE [COMMAND] [ARG...]

Create a new container

Options:
      --add-host list              Add a custom host-to-IP mapping
                                   (host:ip)
  -a, --attach list                Attach to STDIN, STDOUT or STDERR
      --blkio-weight uint16        Block IO (relative weight),
```

图 10-26

下面使用 create 创建容器，镜像是 ubuntu，返回容器 id。通过 ps 可以看到一个状态是 created 的容器，此时的容器并没有启动，如图 10-27 所示。

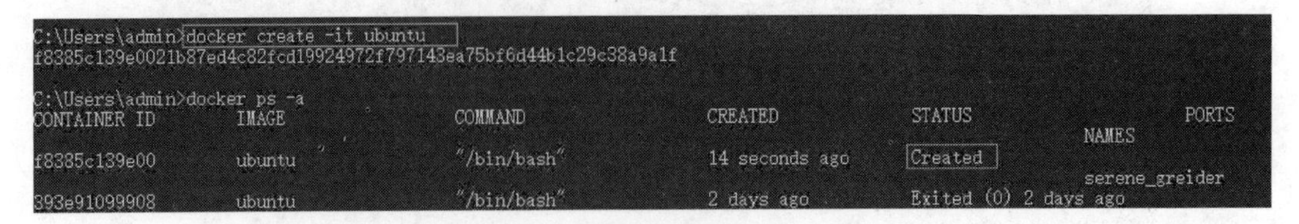

图 10-27

启动容器需要使用 start 命令，还是可以使用 help 来了解一下 start 的使用，如图 10-28 所示。

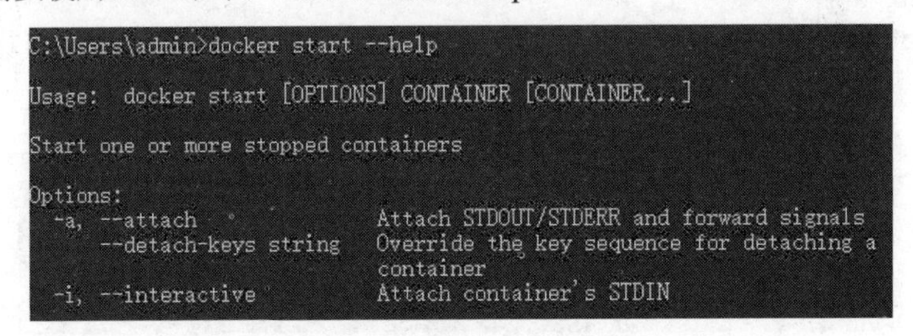

图 10-28

下面使用 start 启动 ubuntu 容器，然后使用 docker ps -a 命令查看 ubuntu，可以发现状态为 Up，如图 10-29 所示。

```
C:\Users\admin>docker start f8385c139e00
f8385c139e00

C:\Users\admin>docker ps -a
CONTAINER ID   IMAGE    COMMAND       CREATED          STATUS                  PORTS   NAMES
f8385c139e00   ubuntu   "/bin/bash"   6 minutes ago    Up 9 seconds                    serene_greider
393e91099908   ubuntu   "/bin/bash"   2 days ago       Exited (0) 2 days ago
```

图 10-29

除了使用 create、start 来启动容器外，还可以使用 run 命令来启动容器，如图 10-30 所示。

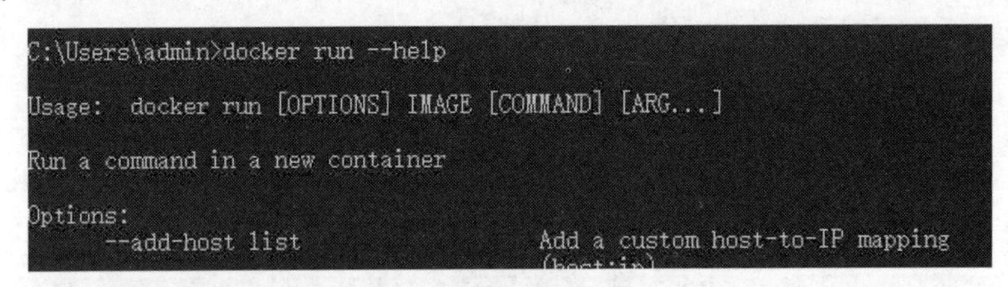

图 10-30

上面是先使用 create 创建一个容器，然后使用 start 启动，其实也可以直接使用 run 来新建并启动容器。当利用 docker run 来创建并启动容器时，Docker 在后台运行的标准操作包括：

（1）检查本地是否存在指定的镜像，不存在就从公有仓库下载。

（2）利用镜像创建一个容器，并启动。

（3）分配一个文件系统给容器，并在只读的镜像层外面挂载一层可读写层。

（4）从宿主主机配置的网桥接口中桥接一个虚拟接口到容器中。

（5）从网桥的地址池配置一个 IP 地址给容器。

（6）执行用户指定的应用程序。

（7）执行完毕后容器自动终止。

图 10-31 所示就是使用 run 命令来启动 ubuntu。

```
C:\Users\admin>docker run -it ubuntu
root@224cc6fed4a1:/#
```

图 10-31

10.3.2 守护态运行

有时候需要让 docker 容器在后台以守护态形式运行，可以通过-d 来实现。如图 10-32 所示，开启一个 ubuntu 容器，启动 bin/sh 下脚本执行后面的脚本语句 docker run -d ubuntu /bin/sh -c "while true;do echo hello world;sleep 10;done;"，让它每隔 10 毫秒输出一个“helloworld”，然后通过 logs 来查看它的输出，可以看到第一次输出了两个，第二次输出了三个。

```
C:\Users\admin>docker run -d ubuntu /bin/sh -c "while true;do echo hello world;sleep 10;done;"
29c19cd0caa73def233fd72035ef38b505c25e772aeb0974bdfed35a901d9c6d

C:\Users\admin>docker logs 29c19cd0caa73def233fd72035ef38b505c25e772aeb0974bdfed35a901d9c6d
hello world
hello world

C:\Users\admin>docker logs 29c19cd0caa73def233fd72035ef38b505c25e772aeb0974bdfed35a901d9c6d
hello world
hello world
hello world
```

图 10-32

10.3.3 终止容器

上面的容器一直在后台运行，可以通过 ps 来查看它的状态，如图 10-33 所示。

```
C:\Users\admin>docker ps -a
CONTAINER ID        IMAGE               COMMAND                  CREATED             STATUS                     PORTS
                                                                                                     NAMES
29c19cd0caa7        ubuntu              "/bin/sh -c 'while t…"   3 minutes ago       Up 3 minutes
                                                                                                     amazing_fermat
21f8a35d349e        ubuntu              "/bin/bash"              6 minutes ago       Exited (0) 6 minutes ago
                                                                                                     focused_stallman
d5d85b454917        ubuntu              "/bin/bash"              7 minutes ago       Exited (0) 7 minutes ago
```

图 10-33

如果想终止它，可以使用 stop。图 10-34 所示是 stop 的格式和参数。

```
C:\Users\admin>docker stop --help

Usage:  docker stop [OPTIONS] CONTAINER [CONTAINER...]

Stop one or more running containers

Options:
  -t, --time int   Seconds to wait for stop before killing it (default 10)
```

图 10-34

使用 stop 命令终止 ubuntu 容器之后，使用 docker ps -a 查看 ubuntu 容器状态，已变成 Exited 状态，如图 10-35 所示。

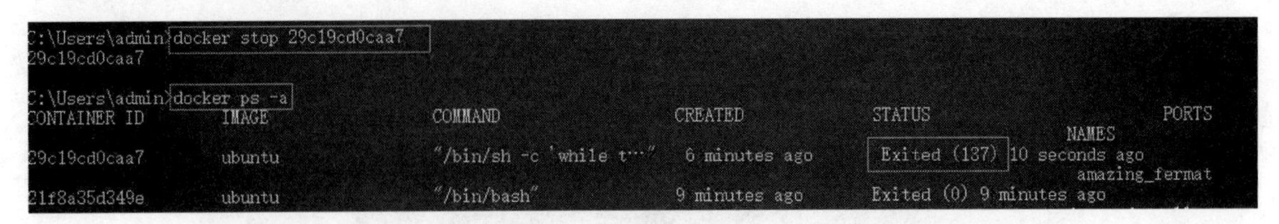
```
C:\Users\admin>docker stop 29c19cd0caa7
29c19cd0caa7

C:\Users\admin>docker ps -a
CONTAINER ID   IMAGE    COMMAND                  CREATED         STATUS                        PORTS   NAMES
29c19cd0caa7   ubuntu   "/bin/sh -c 'while t…"   6 minutes ago   Exited (137) 10 seconds ago           amazing_fermat
21f8a35d349e   ubuntu   "/bin/bash"              9 minutes ago   Exited (0) 9 minutes ago
```

图 10-35

10.3.4 进入容器

启动容器之后就是进入容器，对容器进行操作，包括 attach 命令和 exec 命令。

1. attach 命令

学习 attach 命令还是从 help 命令开始，如图 10-36 所示。

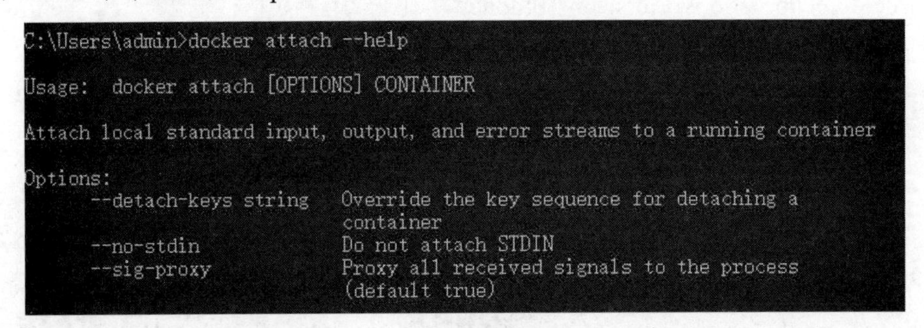
```
C:\Users\admin>docker attach --help

Usage:  docker attach [OPTIONS] CONTAINER

Attach local standard input, output, and error streams to a running container

Options:
      --detach-keys string   Override the key sequence for detaching a
                             container
      --no-stdin             Do not attach STDIN
      --sig-proxy            Proxy all received signals to the process
                             (default true)
```

图 10-36

首先启动 ubuntu，然后使用 attach 命令进入容器。之后使用 echo 输出 abc，如图 10-37 所示。最后使用 Ctrl+P 和 Ctrl+Q 组合键退出。

```
C:\Users\admin>docker run -itd ubuntu
44ce0074a613dac589e575736790fd322158b68a85f24e6fc93dbc2897ee7afd

C:\Users\admin>docker attach keen_golick
root@44ce0074a613:/# echo abc
abc
root@44ce0074a613:/# read escape sequence
```

图 10-37

2. exec 命令

首先使用 help 了解一下 exec 的使用，如图 10-38 所示。

然后使用 exec 命令 docker exec -it keen_golick bin/bash 进入 ubuntu 的 bin/bash 下，输出 abc，如图 10-39 所示。最后使用 Ctrl+P 和 Ctrl+Q 组合键退出。

```
C:\Users\admin>docker exec --help

Usage:  docker exec [OPTIONS] CONTAINER COMMAND [ARG...]

Run a command in a running container

Options:
  -d, --detach               Detached mode: run command in the background
      --detach-keys string   Override the key sequence for detaching a
                             container
  -e, --env list             Set environment variables
  -i, --interactive          Keep STDIN open even if not attached
      --privileged           Give extended privileges to the command
  -t, --tty                  Allocate a pseudo-TTY
  -u, --user string          Username or UID (format:
                             <name|uid>[:<group|gid>])
  -w, --workdir string       Working directory inside the container
```

图 10-38

```
C:\Users\admin>docker exec keen_golick bin/bash

C:\Users\admin>docker exec -it keen_golick bin/bash
root@44ce0074a613:/# echo abc
abc
root@44ce0074a613:/# read escape sequence
```

图 10-39

10.3.5　容器的导入导出

镜像有导入导出，容器也有导入导出。容器的导入导出主要涉及两个命令：export 和 import。

1. 导出

首先还是使用 help 了解一下 export 的使用，如图 10-40 所示。

```
C:\Users\admin>docker export --help

Usage:  docker export [OPTIONS] CONTAINER

Export a container's filesystem as a tar archive

Options:
  -o, --output string   Write to a file, instead of STDOUT
```

图 10-40

使用 docker export -o G:\ubuntu.tar 44ce0074a613 命令将容器 44ce0074a613 导出到 G:\ubuntu.tar 中，之后在 G 盘目录下就会发现 ubuntu.tar，如图 10-41 所示。

图 10-41

2. 导入

导入与导出是相对的，上面介绍了导出，下面介绍一下导入。import 的用法如图 10-42 所示。

```
C:\Users\admin>docker import --help

Usage:  docker import [OPTIONS] file|URL|- [REPOSITORY[:TAG]]

Import the contents from a tarball to create a filesystem image

Options:
  -c, --change list      Apply Dockerfile instruction to the created image
  -m, --message string   Set commit message for imported image
```

图 10-42

这里将上面导出的再导入进来（docker import G:\ubuntu.tar import_ubuntu），如图 10-43 所示。

```
C:\Users\admin>docker import G:\ubuntu.tar import_ubuntu
sha256:e65cfd375bc089b9f47f6a1a173e4db3158eeac5bf39bfcdb6f9286d369da64d

C:\Users\admin>docker images
REPOSITORY            TAG       IMAGE ID       CREATED          SIZE
import_ubuntu         latest    e65cfd375bc0   9 seconds ago    69.8MB
hello-world-import    latest    6fc461936b67   3 days ago       1.85kB
cuiyw-hello-world     latest    1743012f4558   3 days ago       1.84kB
<none>                <none>    002d67520092   3 days ago       1.85kB
```

图 10-43

load 是导入镜像存储文件到本地镜像库，import 是导入一个容器快照到本地镜像库。区别是容器快照文件将丢弃所有的历史记录和元数据信息，仅仅保存容器当前的快照状态。镜像文件将保存完整记录，体积更大，容器快照导入时可以重新指定标签等元数据信息。

10.4 搭建私有仓库

本节学习一下仓库，搭建一个本地私有仓库。当然，也可以使用远程的公有仓库，但在企业中有的还是放在本地，所以需要搭建私有仓库。

1. 搭建仓库

可以在容器中搭建（run）一个仓库镜像，命令为 docker run -d -p 5000:5000 -v /opt/data/registry:/tmp/registry registry，如图 10-44 所示。

```
C:\Users\admin>docker run -d -p 5000:5000 -v /opt/data/registry:/tmp/registry registry
Unable to find image 'registry:latest' locally
latest: Pulling from library/registry
169185f82c45: Pull complete
046e2d030894: Pull complete
188836fddeeb: Pull complete
832744537747: Pull complete
7ceea07e80be: Pull complete
Digest: sha256:870474507964d8e7d8c3b53bcfa738e3356d2747a42adad26d0d81ef4479eb1b
Status: Downloaded newer image for registry:latest
737992080befd5877d6dffbd576a7f9b76411b510b7bc47a3d3d1f499de1257b
```

图 10-44

2. 给待上传镜像设置 tag

给 busybox 设置 tag（docker tag busybox 127.0.0.1:5000/busyboxdocker），然后可以看到 images 中有该 tag 的 image，如图 10-45 所示。

```
C:\Users\admin>docker tag busybox 127.0.0.1:5000/busyboxdocker

C:\Users\admin>docker images
REPOSITORY                     TAG       IMAGE ID       CREATED              SIZE
import_ubuntu                  latest    e65cfd375bc0   About an hour ago    69.8MB
hello-world-import             latest    6fc461936b67   3 days ago           1.85kB
cuiyw-hello-world              latest    1743012f4553   3 days ago           1.84kB
<none>                         <none>    002d67520092   3 days ago           1.85kB
registry                       latest    d0eed8dad114   7 days ago           25.8MB
ubuntu                         latest    20bb25d32758   2 weeks ago          87.5MB
hello-world                    latest    fce289e99eb9   5 weeks ago          1.84kB
127.0.0.1:5000/busyboxdocker   latest    3a093384ac30   5 weeks ago          1.2MB
busybox                        latest    3a093384ac30   5 weeks ago          1.2MB
```

图 10-45

3. 上传到仓库

使用 push 将镜像上传到仓库（docker push 127.0.0.1:5000/busyboxdocker），如图 10-46 所示。

```
C:\Users\admin>docker push 127.0.0.1:5000/busyboxdocker
The push refers to repository [127.0.0.1:5000/busyboxdocker]
683f499823be: Pushed
latest: digest: sha256:bbb143159af9eabdf45511fd5aab4fd2475d4c0e7fd4a5e154b98e838488e510 size: 527
```

图 10-46

10.5 数据管理

在 Docker 的使用过程中，需要对数据进行持久化或需要在多个容器之间进行数据共享，就会涉及容器的数据管理操作。主要有两种方式：一是数据卷；二是数据卷容器。

10.5.1 数据卷

数据卷是一个可供容器使用的特殊目录，将主机操作系统目录直接映射进容器。它可以提供很多特性：

（1）数据卷可以在容器之间共享和重用，容器间传递数据将变得高效方便。

（2）对数据卷内数据的操作会立马生效，无论是容器内还是本地操作。

（3）对数据卷的更新不会影响镜像，解耦了应用与数据。

（4）卷一直存在，直到没有容器使用，可以安全地卸载它。

数据卷可以有两种方式存在：一是挂载主机目录作为数据卷；二是在容器内创建一个数据卷。

1. 挂载主机目录作为数据卷

首先在 G:\docker\test 目录下创建两个文件（1.txt 和 2.log），然后挂载主机目录到数据卷（docker run -d -it --name=test -v G:/docker/test:/test ubuntu /bin/sh），如图 10-47 所示。

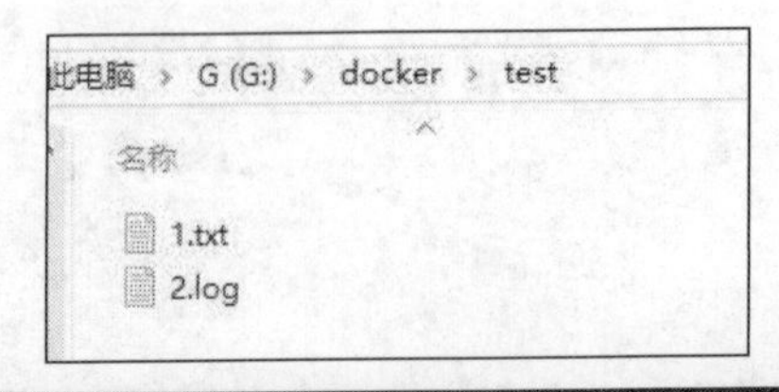


```
C:\Users\admin>docker run -d -it --name=test -v G:/docker/test:/test ubuntu /bin/sh
75ae60cef308618ed479098ed7bf88c1143aabacc93773597ecc031e829bdcc2
```

图 10-47

然后使用 docker ps -a 命令即可找到上面运行的容器，再使用 docker exec -it 75ae60cef308 /bin/sh 命令进入容器打开 bin/sh 目录，会发现有一个 test 的文件夹。打开 test 文件夹，可以看到有上面的两个文件，如图 10-48 所示。

```
C:\Users\admin>docker exec -it 75ae60cef308 /bin/sh
# ls
bin  boot  dev  etc  home  lib  lib64  media  mnt  opt  proc  root  run  sbin  srv  sys  test  tmp  usr  var
# cd test
# ls
1.txt  2.log
#
```

图 10-48

最后在 1.txt 中输入内容，再次打开可以看到容器中的 1.txt 也更新了，如图 10-49 所示。

```
C:\Users\admin>docker exec -it 75ae60cef308 /bin/sh
# ls
bin  boot  dev  etc  home  lib  lib64  media  mnt  opt  proc  root  run  sbin  srv  sys  test  tmp  usr  var
# cd test
# ls
1.txt  2.log
# cat 1.txt
# cat 1.txt
123abc#
```

图 10-49

2. 在容器内创建一个数据卷

上面是挂载主机目录作为数据卷，其实也可以在容器内创建一个数据卷，使用 docker run -d -it --name=vtest -v /vtest ubuntu /bin/sh 命令，创建了 ubuntu 镜像的容器 vtest，并在容器中创建了一个名为 vtest 的数据卷，如图 10-50 所示。

```
C:\Users\admin>docker run -d -it --name=vtest -v /vtest ubuntu /bin/sh
9664652a8a47eb83d26cbeb6df8af503d943ef0df68ddad56faee6fc8640e1bb
```

图 10-50

使用 exec 命令进入容器，可以发现定义的 vtest 数据卷，如图 10-51 所示。

```
C:\Users\admin>docker exec -it 9664652a8a47  /bin/sh
# ls
bin  boot  dev  etc  home  lib  lib64  media  mnt  opt  proc  root  run  sbin  srv  sys  tmp  usr  var  vtest
#
```

图 10-51

10.5.2　数据卷容器

如果容器之间需要共享一些持续更新的数据，最简单的方式就是用户数据卷容器。数据卷容器就是一种普通容器，专门提供数据卷供其他容器挂载使用。

我们在上面通过挂载主机目录作为数据卷创建了 test 容器，这里可以把 test 容器作为数据卷容器，然后通过下面的命令另外创建了 db1、db2 两个容器，使用--volumes-from 参数挂载 test 容器中的数据卷。

```
docker run -it --volumes-from test --name db1 ubuntu /bin/sh
docker run -it --volumes-from test --name db2 ubuntu /bin/sh
```

之后依次使用 exec 命令打开 db1、db2 两个容器，在两个容器下都能找到 test 目录下的 1.txt 和 2.log 文件，在容器 db2 中打开 1.txt 并显示其内容，之后修改 1.txt 文件的内容，再打开容器 db1 中的 1.txt，可以发现内容已更改最新，如图 10-52 所示。

```
docker exec -it 0abd9b51c263 /bin/sh
docker exec -it 4e76041f9861 /bin/sh
```

```
C:\Users\admin>docker ps -a
CONTAINER ID   IMAGE    COMMAND        CREATED             STATUS                      PORTS   NAMES
0abd9b51c263   ubuntu   "/bin/sh"      7 minutes ago       Up 7 minutes                        db2
4e76041f9861   ubuntu   "/bin/sh"      8 minutes ago       Up 8 minutes                        db1
e44cd1b506b0   ubuntu   "/bin/sh"      17 minutes ago      Up 16 minutes                       test
330c844f3925   ubuntu   "/bin/bash"    About an hour ago   Exited (255) 41 minutes ago         wizardly_franklin

C:\Users\admin>docker exec -it 0abd9b51c263 /bin/sh
# ls
bin  boot  dev  etc  home  lib  lib64  media  mnt  opt  proc  root  run  sbin  srv  sys  test  tmp  usr  var
# cd test
# ls
1.txt  2.log
# cat 1.txt
123abc# read escape sequence

C:\Users\admin>docker exec -it 4e76041f9861 /bin/sh
# ls
bin  boot  dev  etc  home  lib  lib64  media  mnt  opt  proc  root  run  sbin  srv  sys  test  tmp  usr  var
# cd test
# ls
1.txt  2.log
# cat 1.txt
123abcd#
```

图 10-52

10.6　端口映射与容器互联

在搭建私有仓库时，用到的命令为 docker run -d -p 5000:5000 -v /opt/data/registry:/tmp/registry registry，其中“-p 5000:5000”的作用就是用来进行端口映射。那为什么需要端口映射呢？其实在启动容器时，如果不配置宿主机器与虚拟机的端口映射，外部程序是无法访问虚拟机的，因为没有端口，所以需要进行端口映射。

10.6.1 端口映射

端口映射有两个关键词：-P 和-p（见图 10-53），一个是大写，一个是小写，通过 run --help 可以看到。大写的 P 是随机映射一个 49000~49900 的端口到内部容器开放的网络端口。小写 p 可以指定要映射的端口，并且在一个指定端口上只可以绑定一个容器。

```
-p, --publish list                Publish a container's port(s) to
                                  the host
-P, --publish-all                 Publish all exposed ports to
                                  random ports
    --read-only                   Mount the container's root
                                  filesystem as read only
    --restart string              Restart policy to apply when a
                                  container exits (default "no")
    --rm                          Automatically remove the container
                                  when it exits
    --runtime string              Runtime to use for this container
    --security-opt list           Security Options
    --shm-size bytes              Size of /dev/shm
    --sig-proxy                   Proxy received signals to the
                                  process (default true)
    --stop-signal string          Signal to stop a container
                                  (default "15")
    --stop-timeout int            Timeout (in seconds) to stop a
                                  container
    --storage-opt list            Storage driver options for the
                                  container
    --sysctl map                  Sysctl options (default map[])
    --tmpfs list                  Mount a tmpfs directory
```

图 10-53

支持的格式有如下三种，比较常用的是第三种，因为这样没有对 ip 进行限制，移植也方便。

（1）ip:hostPort:containerPort：映射指定地址的指定端口到虚拟机的指定端口（不常用），如 127.0.0.1:5000:5000，映射本机的 5000 端口到虚拟机的 5000 端口。

（2）ip::containerPort：映射指定地址的任意端口到虚拟机的指定端口（不常用），如 127.0.0.1::5000，映射本机的 5000 端口到虚拟机的 5000 端口。

（3）hostPort:containerPort：映射本机的指定端口到虚拟机的指定端口（常用），如 5000:5000，映射本机的 5000 端口到虚拟机的 5000 端口。

映射完之后我们可以通过 ps 来查看容器的映射情况，这里就不再演示了。其实也可以使用 port 命令来查看映射情况。下面先通过 help 来查看 port 的语法和格式，然后查看容器 peaceful_lovelace 的端口映射情况，如图 10-54 所示。另外，可以多次使用-p 来绑定多个端口。

```
C:\Users\admin>docker port --help

Usage:  docker port CONTAINER [PRIVATE_PORT[/PROTO]]

List port mappings or a specific mapping for the container

C:\Users\admin>docker port 884a068364ea
5000/tcp -> 0.0.0.0:5000
```

图 10-54

10.6.1 容器互联

在实际应用中往往需要多个容器交互，比如一个数据库容器来提供 db 服务，多个应用容器来部署应用，使用端口访问就会暴露端口，这样不太安全。有没有方法让容器互联呢？答案当然是有。容器互联的方法有多种，这里主要学习 link，但 link 是只针对单宿主主机的。

（1）启动 mysql server

首先使用 run 实例了一个 mysql 数据库的容器，容器名为 mysql，123 为数据库的 root 密码。

```
docker run --name mysql -e MYSQL_ROOT_PASSWORD=123 -d mysql
```

（2）使用--link 关联

这里启动一个 webapp 的容器，关联 mysql 容器，并取别名为 db。webapp 容器的镜像就是普通的 ubuntu 镜像，如图 10-55 所示。

```
docker run --name webapp --link mysql:db -it -d ubuntu /bin/sh
```

```
C:\Users\admin>docker run --name webapp --link mysql:db -it -d ubuntu /bin/sh
61b2a72ae398589526aa0530ffc50fcb3f6497f0d1c35250c88b644128034b2f
```

图 10-55

两个容器互联相当于在它们之间创建了一个虚拟通道，而且不用映射它们的端口到宿主主机上。通过两种方式为容器公开连接信息：一是更新环境变量；二是更新/etc/hosts 文件。

我们可以进入容器 webapp 来查看它的环境变量和 hosts 文件。可以在 hosts 中找到 172.17.0.6 db 40cb0c95f62d mysql 配置信息，在环境变量中也能找到 MySQL 数据库的相关配置信息，如图 10-56 所示。

```
C:\Users\admin>docker exec -it 61b2a72ae398 /bin/sh
# cat /etc/hosts
127.0.0.1       localhost
::1     localhost ip6-localhost ip6-loopback
fe00::0 ip6-localnet
ff00::0 ip6-mcastprefix
ff02::1 ip6-allnodes
ff02::2 ip6-allrouters
172.17.0.6      db 40cb0c95f62d mysql
172.17.0.7      61b2a72ae398
# env
DB_PORT_33060_TCP_ADDR=172.17.0.6
HOSTNAME=61b2a72ae398
DB_PORT=tcp://172.17.0.6:3306
DB_PORT_3306_TCP=tcp://172.17.0.6:3306
DB_PORT_33060_TCP_PORT=33060
DB_ENV_MYSQL_VERSION=8.0.15-1debian9
DB_PORT_33060_TCP_PROTO=tcp
HOME=/root
DB_NAME=/webapp/db
DB_PORT_33060_TCP=tcp://172.17.0.6:33060
TERM=xterm
PATH=/usr/local/sbin:/usr/local/bin:/usr/sbin:/usr/bin:/sbin:/bin
DB_ENV_MYSQL_MAJOR=8.0
DB_PORT_3306_TCP_ADDR=172.17.0.6
DB_ENV_MYSQL_ROOT_PASSWORD=123
PWD=/
DB_ENV_GOSU_VERSION=1.7
DB_PORT_3306_TCP_PORT=3306
DB_PORT_3306_TCP_PROTO=tcp
#
```

图 10-56

10.7 Dockerfile

Dockerfile 是一个包含用于组合映像的命令的文本文档。可以在 docker build 命令中使用-f 标志指向文件系统中任何位置的 Dockerfile：docker build -f /path/to/a/Dockerfile。

Docker 通过读取 Dockerfile 中的指令自动生成映像。它一般分为四部分：基础镜像信息、维护者信息、镜像操作指令和容器启动时执行指令，‘#’为 Dockerfile 中的注释。在这四部分中包含若干个指令，下面是常用指令的介绍。

（1）FROM 基础镜像信息

指定所创建镜像的基础镜像，必须指定且需要在 Dockerfile 其他指令的前面。后续的指令都依赖于该指令指定的 image。FROM 指令指定的基础 image 可以是官方远程仓库中的，也可以位于本地仓库，格式为 FROM <image>或 FROM <image>:<tag>。

（2）MAINTAINER 指定维护者信息

用于将 image 的制作者相关的信息写入到 image 中。当我们对该 image 执行 docker inspect 命令时，输出中有相应的字段记录该信息，格式为 MAINTAINER <name>。

（3）RUN 运行指定命令

格式为 RUN<command>或 RUN["executable","param1","param2"]。前者默认在 shell 终端中运行命令，即/bin/sh -c；后者则使用 exec 执行，不会启动 shell 环境，指定使用其他终端类型可以通过第二种方式实现，例如 RUN ["/bin/bash","-c","each hello"]。每条 RUN 指令将在当前镜像的基础上执行指定命令，并提交为新的镜像，当命令较长时可以使用\来换行。

（4）CMD

用于 container 启动时指定的操作。该操作可以是执行自定义脚本，也可以是执行系统命令。该指令只能在文件中存在一次，如果有多个，则只执行最后一条。格式为 CMD["executable","param1","param2"]。使用 exec 执行。CMD command param1 param2 在/bin/sh 中执行，提供给需要交互的应用。CMD ["param1","param2"]提供给 ENTRYPORT 的默认参数。

（5）ENTRYPOINT

指定容器启动时执行的命令，可以多次设置，但是只有最后一个有效。

```
ENTRYPOINT ["executable", "param1", "param2"] (like an exec, the preferred
form)
ENTRYPOINT command param1 param2 (as a shell)
```

该指令的使用分为两种情况，一种是独自使用；另一种和 CMD 指令配合使用。当独自使用时，如果你还使用了 CMD 命令且 CMD 是一个完整的可执行的命令，那么 CMD 指令和 ENTRYPOINT 会互相覆盖，只有最后一个 CMD 或者 ENTRYPOINT 有效。另一种用法和 CMD 指令配合使用来指定 ENTRYPOINT 的默认参数，这时 CMD 指令不是一个完整的可执行命令，仅仅是参数部分；ENTRYPOINT 指令只能使用 JSON 方式指定执行命令，而不能指定参数。

（6）LABEL

给镜像添加信息。格式为 LABEL <key>=<value>。

（7）EXPOSE

该指令会将容器中的端口映射成宿主机器中的某个端口。当你需要访问容器的时候，可以不使用容器的 IP 地址而是使用宿主机器的 IP 地址和映射后的端口。要完成整个操作，需要两个步骤，首先在 Dockerfile 使用 EXPOSE 设置需要映射的容器端口，然后在运行容器的时候指定-p 选项加上 EXPOSE 设置的端口，这样 EXPOSE 设置的端口号会被随机映射成宿主机器中的一个端口号。也可以指定需要映射到宿主机器的那个端口，这时要确保宿主机器上的端口号没有被使用。EXPOSE 指令可以一次设置多个端口号，相应的运行容器的时候，可以配套地多次使用-p 选项。

（8）ENV

指定环境变量，格式为 ENV<key><value>或 ENV <key>=<value>，在镜像生成过程中会被后续的 RUN 指令使用，在镜像启动的容器中也会存在。也可以通过 docker run --env key=value 设置或修改环境变量。假如你安装了 Java 程序，需要设置 JAVA_HOME，那么可以在 Dockerfile 中这样写：ENV JAVA_HOME /path/to/java/dirent。

（9）ADD

所有复制到 container 中的文件和文件夹权限为 0755，uid 和 gid 为 0；如果是一个目录，那么会将该目录下的所有文件添加到 container 中，不包括目录；如果文件是可识别的压缩格式，则 docker 会帮忙解压缩（注意压缩格式）；如果<src>是文件且<dest>中不使用斜杠结束，则会将<dest>视为文件，<src>的内容会写入<dest>；如果<src>是文件且<dest>中使用斜杠结束，则会将<src>文件复制到<dest>目录下。格式为 ADD <src> <dest>。其中，<src>是相对被构建的源目录的相对路径，可以是文件或目录的路径，也可以是一个远程的文件 url；<dest> 是 container 中的绝对路径。

（10）COPY

复制本地主机的 src（为 Dockerfile 所在目录的相对路径、文件或目录）下的内容到镜像中的 dest 下，目标路径不存在时会自动创建，格式为 COPY <src> <dest>。

（11）VOLUME

使容器中的一个目录具有持久化存储数据的功能，该目录可以被容器本身使用，也可以共享给其他容器使用。我们知道容器使用的是 AUFS，这种文件系统不能持久化数据，当容器关闭后，所有的更改都会丢失。当容器中的应用有持久化数据的需求时可以在 Dockerfile 中使用该指令，格式为 VOLUME ["/data"]。

（12）USER

设置启动容器的用户，默认是 root 用户。格式为 USER daemon。当服务不需要管理员权限时，可以通过该命令指定运行用户。

（13）WORKDIR

为后续的 RUN CMD ENTRYPOINT 指定配置工作目录，格式为 WORKDIR /path/to/workdir。

（14）ONBUILD

配置当所创建的镜像作为其他镜像的基础镜像时所执行的创建操作指令，格式为 ONBUILD [INSTRUCTION]。

（15）ARG

指定一些镜像内使用的参数（例如版本号信息等）。这些参数在执行 docker build 命令时以 --build-arg <varname>=<value>格式传入。格式为 ARG <name>[=default value]。可以用 docker build --build-arg <name>=<value>来指定参数值。

10.8 Docker 容器 Tomcat 部署 war 包

前面几节介绍了 Docker 的基础知识，本节是前面章节的具体应用，在 Docker 容器中，使用 Tomcat 部署 war 包主要包括四个步骤，即创建 Tomcat 容器、上传 war 包到容器、重启容器、访问应用。

1. 创建 Tomcat 容器

使用 docker run -d --name cmdtomcat -p 8080:8080 tomcat 命令启动 Tomcat 容器，在浏览器中输入 http://localhost:8080/可以显示 tomcat 配置页面。使用 docker ps -a 可以看到 cmdtomcat 的容器，如图 10-57 所示。

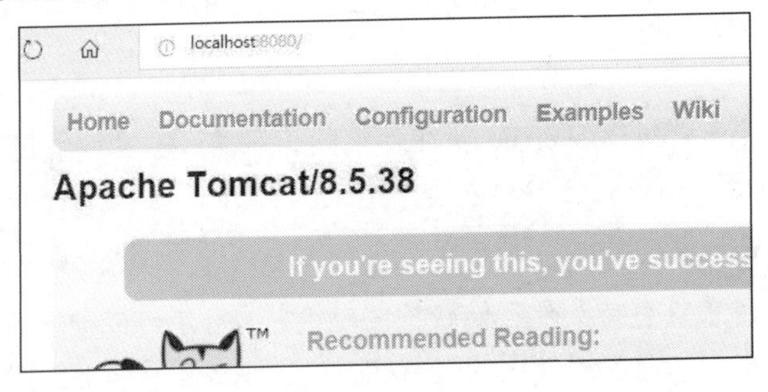

图 10-57

使用 docker exec -it 6ba4ccc864bd /bin/bash 命令交互式地进入 Tomcat 容器，可以看到 webapps 目录以及 webapps 目录下的文件，如图 10-58 所示。

2. 上传 war 包到 Tomcat 容器

war 包是第 9 章打包部署用的包，为了方便演示，这里将 war 包放在了 G:\dockerwar 目录下。Docker 中也可以使用 cp 命令完成和宿主机的文件复制，如图 10-59 所示。

```
C:\Users\admin>docker exec -it 6ba4ccc864bd /bin/bash
root@6ba4ccc864bd:/usr/local/tomcat# ls
BUILDING.txt  CONTRIBUTING.md  LICENSE  NOTICE  README.md  RELEASE-NOTES  RUNNING.t
xt  bin  conf  include  lib  logs  native-jni-lib  temp  webapps  work
root@6ba4ccc864bd:/usr/local/tomcat# cd webapps
root@6ba4ccc864bd:/usr/local/tomcat/webapps# ls
ROOT  docs  examples  host-manager  manager
root@6ba4ccc864bd:/usr/local/tomcat/webapps#
```

图 10-58

```
Docker cp G:\dockerwar\SpringBootSwagger-0.0.1-SNAPSHOT.war
cmdtomcat:/usr/local/tomcat/webapps
```

```
C:\Users\admin>docker cp  G:\dockerwar\SpringBootSwagger-0.0.1-SNAPSHOT.war
cmdtomcat:/usr/local/tomcat/webapps
```

图 10-59

3. 重启容器

使用 docker restart cmdtomcat 命令重启容器，然后进入容器可以看到上面复制进去的 SpringBootSwagger-0.0.1-SNAPSHOT.war 包，如图 10-60 所示。

```
C:\Users\admin>docker restart cmdtomcat
cmdtomcat

C:\Users\admin>docker exec -it cmdtomcat /bin/bash
root@6ba4ccc864bd:/usr/local/tomcat# ls
BUILDING.txt  CONTRIBUTING.md  LICENSE  NOTICE  README.md  RELEASE-NOTES  RUN
NING.txt  bin  conf  include  lib  logs  native-jni-lib  temp  webapps  work
root@6ba4ccc864bd:/usr/local/tomcat# cd webapps
root@6ba4ccc864bd:/usr/local/tomcat/webapps# ls
ROOT  SpringBootSwagger-0.0.1-SNAPSHOT  SpringBootSwagger-0.0.1-SNAPSHOT.war
 docs  examples  host-manager  manager
root@6ba4ccc864bd:/usr/local/tomcat/webapps#
```

图 10-60

4. 访问应用

在浏览器中输入“http://localhost:8080/SpringBootSwagger-0.0.1-SNAPSHOT/swagger-ui.html”即可正常显示页面，如图 10-61 所示。

图 10-61

5. 以挂载方式启动

上述执行有一个弊端，就是容器重启后项目就会不在了，我们可以以挂载的方式启动，如图 10-62 所示。

```
C:\Users\admin>docker run -d -v G:/dockerwar/SpringBootSwagger-0.0.1-SNAPSHO
T.war:/usr/local/tomcat/webapps/SpringBootSwagger-0.0.1-SNAPSHOT.war --name
cmdtomcat2 -p 8081:8080 tomcat
6004e6f509f65076545590d96c231a4074047a79c32077052e64ea6640c550d3
```

图 10-62

如图 10-63 所示，进入 cmdtomcat2 容器，可以看到 SpringBootSwagger-0.0.1-SNAPSHOT.war 已在 webapps 目录下。在浏览器中输入“http://localhost:8081/SpringBootSwagger-0.0.1-SNAPSHOT/swagger-ui.html”，也是正常显示页面。

```
C:\Users\admin>docker exec -it cmdtomcat2 /bin/bash
root@6004e6f509f6:/usr/local/tomcat# ls
BUILDING.txt  CONTRIBUTING.md  LICENSE  NOTICE  README.md  RELEASE-NOTES  RU
NNING.txt  bin  conf  include  lib  logs  native-jni-lib  temp  webapps  wor
k
root@6004e6f509f6:/usr/local/tomcat# cd webappps
bash: cd: webappps: No such file or directory
root@6004e6f509f6:/usr/local/tomcat# cd webapps
root@6004e6f509f6:/usr/local/tomcat/webapps# ls
ROOT  SpringBootSwagger-0.0.1-SNAPSHOT  SpringBootSwagger-0.0.1-SNAPSHOT.war
  docs  examples  host-manager  manager
root@6004e6f509f6:/usr/local/tomcat/webapps#
```

图 10-63

10.9 小结

本章主要学习了 Docker 的基础知识，了解了 Docker 的三大核心概念：镜像、容器、仓库，通过使用 Tomcat 部署 war 演示了 Docker 的简单使用。如果对 Docker 感兴趣，后续还可以学习 K8s、Docker Compose。

第 11 章

用户权限管理项目实战

前面几章学习了 Spring、MyBatis、Spring Boot 等常用的相关技术，知识点比较分散，本章通过项目实战来对前面几个章节知识点进行回顾，同时也方便初学者了解实际项目开发流程。

本章主要涉及的知识点：

- 项目基础：需求分析、RBAC 介绍。
- 项目技术选型：技术方案选定、项目框架搭建。
- 项目技术实现：数据库设计、功能点详细设计和具体实现。
- Shiro 框架：使用 Shiro 实现权限管理。

11.1 项目基础

在软件开发过程中，一般都会经历需求分析、可行性分析、概要设计、详细设计、编码、测试、交付、验收、维护等阶段，首先要了解项目基本信息、需要解决哪些问题、需求有哪些，然后对需求进行可行性研究，再根据确定的需求选用合适的解决方案。

11.1.1 项目介绍

系统一般由多个模块组成，用户权限管理算是一个必不可少的模块，而且不管是什么系统一般都会有该模块。用户权限管理是使用最广的也是最基础、最重要的模块。不良的权限管理系统，必然留下系统漏洞，给黑客可趁之机。很多软件可以轻松通过 URL 侵入、SQL 注入等模式，轻松越权获得未授权数据，甚至对系统数据进行修改、删除，造成巨大损失。

其实实现该项目的目的不仅是完成用户权限管理模块，而是在过程中了解软件开发流程和设计思想，起到抛砖引玉的作用。系统是由多个模块组成的，熟悉了一个模块的开发流程，掌握了软件开发设计思想，其他模块也是按部就班地实现。

11.1.2 需求分析

以公司为例，一个公司有产品、开发、运维之分，各自负责各自的业务，相互独立，又相互协作，共同完成一个任务，拥有不同权限的用户查看不同的页面，进行不同的操作。拥有不同权限的用户查看不同的页面、进行不同的操作，其实就是需求，下面就是对该需求的分析。

首先分析拥有不同权限的用户部分，有的用户权限很少，有的用户权限很多，而且用户的权限也会发生变动，用户的数量也会发生变化，有时用户数量甚至会呈几何级增长，如果对每个用户都专门进行权限判断，那对系统维护和权限管理都带有很大挑战。拥有不同权限的原因是岗位分工不同，那我们可以通过岗位来控制权限，不同岗位的用户有不同的权限，相同岗位的用户有相同的权限，这里岗位也可以理解成角色。所以功能权限管理技术，一般就使用基于角色访问控制技术（Role Based Access Control，RBAC）。

在 RBAC 中，权限与角色相关联，用户通过成为适当角色的成员而得到这些角色的权限。这就极大地简化了权限的管理。这样管理都是层级相互依赖的，权限赋予角色，而把角色又赋予用户，这样的权限设计很清楚，管理起来很方便。该技术模型如图 11-1 所示。

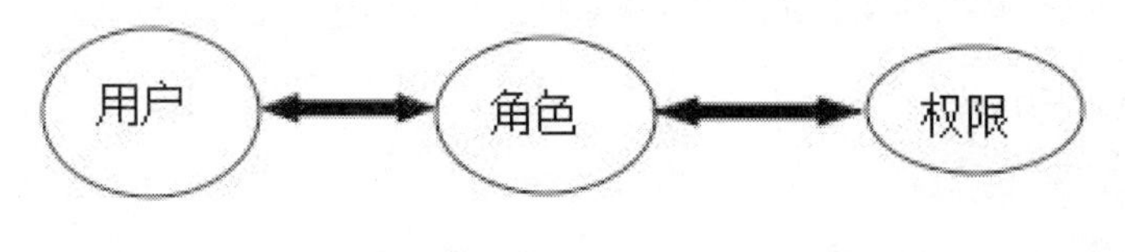

图 11-1

然后分析查看不同的页面，进行不同的操作部分。首先对不同页面的理解，要弄清楚是怎么不同，比如有的用户对一些页面完全不能访问，有的用户可以访问，不同用户访问同一页面显示的数据、内容不同。其次是不同的操作的理解，同一页面的按钮，有的用户可以点击，有的用户无法点击，有的用户有编辑数据的权限，有的用户没有编辑数据的权限。这里要特别提醒一下，需求分析是非常重要的，调研需求的方法也多种多样，只有先把需求理清、明确，知道要解决什么问题了才好进行后续的设计开发工作，需求不明也是好多项目延期、加班严重的一个重要原因。

11.1.3 技术选型

项目主要使用 LayUI 前端框架和 Thymeleaf，实现前后端分离，后端主要使用 Spring Boot、Spring MVC、MyBatis、PageHelper，数据库使用 MySQL，日志使用 Log4j2，权限管理使用 Shiro。

对于前端框架有好多种选择：Bootstrap、EasyUI、KendoUI、LayUI 等各种前端框架。选择哪一种都可以，这里选择 LayUI 框架。

后端几个技术的选择，主要是巩固一下前面章节的知识，对前面章节的内容进行复习，也是目前项目比较常用的技术。

权限管理选择的是 Shiro 框架，是 Java 的一个安全框架。目前，使用 Apache Shiro 的人越来越多，因为它相当简单，对比 Spring Security，可能没有 Spring Security 做的功能强大，但是在实际工作时可能并不需要那么复杂的东西，所以使用小而简单的 Shiro 就足够了。对于 Shiro 的详细介绍，大家可以从网络上查看。

11.2 项目实现

前面对项目进行了分析以及技术选型，本节就进入项目实现部分。首先，需要把项目框架搭建起来。

11.2.1 搭建框架

本项目的架构采用分层思想，主要包含 8 个包，如图 11-2 所示。

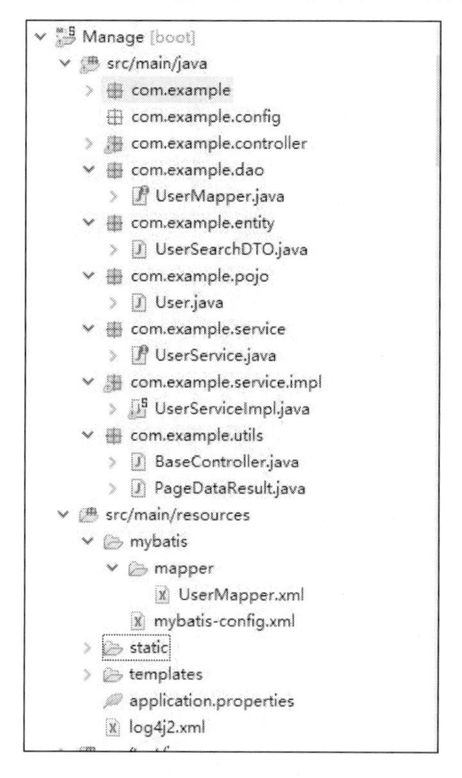

图 11-2

下面对这几个包进行简单介绍。

- com.example：存在 main 函数类。
- com.example.config：配置类包，例如 druid 多数据源配置类。
- com.example.controller：存放 controller。
- com.example.dao：与数据库交互层，存放 mapper 接口。
- com.example.entity：实体层，这个与 com.example.pojo 有点类似，不过两个还是有区别的，pojo 层主要是与数据库单个数据表对应，entity 可能是其他对象的抽象，比如多个表联合查询组成的行对应的类或者前端页面显示对应的类、一对多、多对多关系。
- com.example.pojo：数据库表的对应类。

- com.example.service：业务服务接口层，定义服务接口。优势是什么呢？这样在 Controller 中注入的是 service 接口，是面向接口的编程，如果服务的具体实现改变了也不影响其他注入类。
- com.example.service.impl：实现服务接口，业务逻辑的具体实现。
- com.example.utils：基础类、工具类层。

静态文件或者资源文件的目录结构如图 11-3 所示。其中，mybatis 目录下存放 mapper 文件，static 目录中存放 css、js 等资源文件，templates 中存在 html 页面。

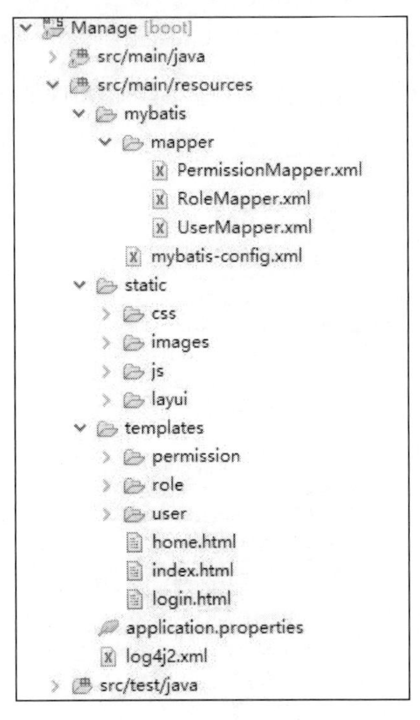

图 11-3

11.2.2 数据库设计

基于前面 RBAC 的分析，设计了 3 张实体表（用户表 user、角色 role、权限 permission）和 2 张关系表（用户角色关系表 user_role、角色权限关系 role_permission）。下面是创建几张表的 SQL 语句。

User 表：

```
CREATE TABLE `user` (
  `name` varchar(20) DEFAULT NULL,
  `age` int(11) DEFAULT NULL,
  `sex` tinyint(1) DEFAULT '0',
  `pwd` varchar(45) DEFAULT '123456',
  `id` varchar(45) NOT NULL,
  PRIMARY KEY (`id`)
) ENGINE=InnoDB DEFAULT CHARSET=utf8;
```

Role 表：

```
CREATE TABLE `role` (
  `name` varchar(20) DEFAULT NULL,
  `desp` varchar(50) DEFAULT NULL,
  `id` varchar(20) NOT NULL,
  PRIMARY KEY (`id`)
) ENGINE=InnoDB DEFAULT CHARSET=utf8;
```

Permission 表：

```
CREATE TABLE `permission` (
  `id` varchar(45) NOT NULL,
  `name` varchar(45) DEFAULT NULL,
  `code` varchar(45) DEFAULT NULL,
  `desp` varchar(45) DEFAULT NULL,
  `url` varchar(45) DEFAULT NULL,
  PRIMARY KEY (`id`)
) ENGINE=InnoDB DEFAULT CHARSET=utf8;
```

User_Role 表：

```
CREATE TABLE `user_role` (
  `userid` varchar(45) NOT NULL,
  `roleid` varchar(45) NOT NULL,
  PRIMARY KEY (`userid`,`roleid`)
) ENGINE=InnoDB DEFAULT CHARSET=utf8;
```

Role_Permission 表：

```
CREATE TABLE `role_permission` (
  `roleid` varchar(45) NOT NULL,
  `permid` varchar(45) NOT NULL,
  PRIMARY KEY (`roleid`,`permid`)
) ENGINE=InnoDB DEFAULT CHARSET=utf8;
```

11.2.3　前端框架引入

前端选择的是 LayUI 框架，项目首页采用常见的左侧导航栏，右侧上部是 tab 组件，下部显示内容，这里推荐一个基于开源 LayUI 的插件 layTabPlus。该插件是一个 layUI 后台 Tab 布局框架的扩展插件，实现了 Tab 管理、刷新按钮、iframe 优化等功能，非常好用。引入方法也比较简单。

1. 下载插件

需要下载两个插件：LayUI 插件和 layTabPlus 插件。layTabPlus 插件的下载地址为 https://gitee.com/Leytton/layTabPlus。两个插件放到项目的资源包 resources 下，把 layTabPlus 插件中的 index.html 页面内容放到 templates 下，这里 index.html 用作了欢迎页面，所以使用了 home.html。

2. 引入 thymeleaf 模块

新建项目时选择 Web 和 thymeleaf 模块，或者手动在项目的 pom.xml 中引入 spring-boot-starter-thymeleaf 来自动集成 thymeleaf。

3. 创建 Controller

在 com.example.Controller 包下创建 HomeController，用来测试 home.html。

```
package com.example.controller;

import org.springframework.stereotype.Controller;
import org.springframework.ui.Model;
import org.springframework.web.bind.annotation.RequestMapping;
import org.springframework.web.bind.annotation.RequestMethod;

@Controller
@RequestMapping("/home")
public class HomeController {
    @RequestMapping(value = "/home",method = RequestMethod.GET)
    public String hello(Model model) {
        return "home";
    }
}
```

4. 测试

启动项目，在浏览器中输入“http://localhost:8080/home/home”，如果显示如图 11-4 所示的页面，LayUI 框架就引入项目中。

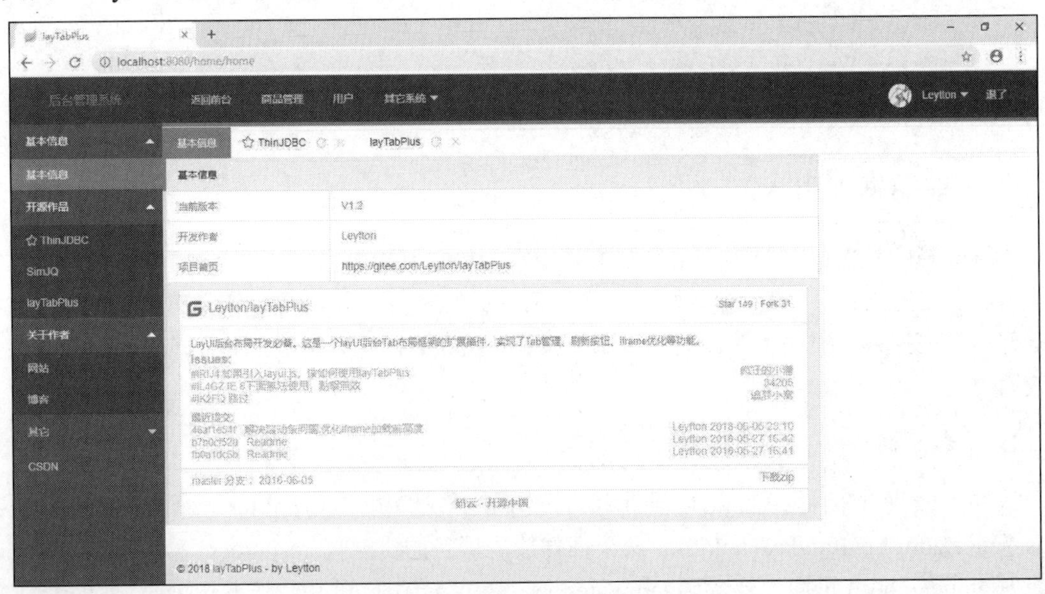

图 11-4

11.2.4 用户角色增删改查

1. 引入依赖

上一小节引入了前端框架 LayUI 和 thymeleaf 实现前后端分离，这里实现用户角色增、删、改、查功能，需要对数据库操作，所以在项目中引入了 mybatis-spring-boot-starter、mysql-connector-java，并引入了分页插件 pagehelper-spring-boot-starter 和日志模块 spring-boot-starter-log4j2。

```
    <!-- https://mvnrepository.com/artifact/org.mybatis.spring.boot/
mybatis-spring-boot-starter -->
    <dependency>
        <groupId>org.mybatis.spring.boot</groupId>
        <artifactId>mybatis-spring-boot-starter</artifactId>
        <version>1.3.2</version>
    </dependency>
    <!-- https://mvnrepository.com/artifact/mysql/mysql-connector-java -->
    <dependency>
        <groupId>mysql</groupId>
        <artifactId>mysql-connector-java</artifactId>
        <version>8.0.11</version>
    </dependency>
    <!-- https://mvnrepository.com/artifact/com.github.pagehelper/
pagehelper-spring-boot-starter -->
    <dependency>
        <groupId>com.github.pagehelper</groupId>
        <artifactId>pagehelper-spring-boot-starter</artifactId>
        <version>1.2.10</version>
    </dependency>
    <dependency>
        <groupId>org.springframework.boot</groupId>
        <artifactId>spring-boot-starter-log4j2</artifactId>
    </dependency>
```

集成 log4j2 后可以在 application.properties 中设置 log4j2 的相关配置或者创建 log4j2.xml 后放在 application.properties 目录下。这里在创建的 log4j2.xml 中设置日志存放位置为 D:\log\logs。在该目录下的日志文件如图 11-5 所示。

电脑 › 软件 (D:) › log › logs

名称	修改日期	类型	大小
2019-02	2019/2/26 0:08	文件夹	
manage-error.log	2019/2/26 0:13	文本文档	39 KB
manage-info.log	2019/2/26 0:42	文本文档	74 KB
manage-warn.log	2019/2/25 23:43	文本文档	17 KB

图 11-5

在引入 log4j2 时还需要将 Spring Boot 默认的日志模块排除，需要在 spring-boot-starter-web 中增加 exclusions 子元素。

```
<dependency>
   <groupId>org.springframework.boot</groupId>
   <artifactId>spring-boot-starter-web</artifactId>
   <exclusions>
       <exclusion>
           <artifactId>spring-boot-starter-logging</artifactId>
           <groupId>org.springframework.boot</groupId>
       </exclusion>
   </exclusions>
</dependency>
```

2. 单表增/删/改/查

在用户管理页面，实现了用户的增加、删除、修改、查询和批量删除，首先是定义实体类（在 POJO 层定义 User 实体类）。

```
package com.example.pojo;
import java.io.Serializable;
import java.util.List;

public class User implements Serializable{

   private static final long serialVersionUID = 1L;
   private String id;
   private String name;
   private Integer age;
   private String pwd;
   private String sex;
   private List<Role> roles;

   public String getId() {
       return id;
   }
   public void setId(String id) {
       this.id = id;
   }
   public String getName() {
       return name;
   }
   public void setName(String name) {
       this.name = name;
   }
   public Integer getAge() {
       return age;
   }
```

```
    public void setAge(Integer age) {
        this.age = age;
    }
    public String getPwd() {
        return pwd;
    }
    public void setPwd(String pwd) {
        this.pwd = pwd;
    }

    public List<Role> getRoles() {
        return roles;
    }
    public void setRoles(List<Role> roles) {
        this.roles = roles;
    }
    @Override
    public String toString() {
        return "User [id=" + id + ", name=" + name + ", age=" + age + ", pwd="
+ pwd + ", sex=" + sex + ", roles="
                + roles + "]";
    }
    public String getSex() {
        return sex;
    }
    public void setSex(String sex) {
        this.sex = sex;
    }
}
```

然后是在 DAO 层定义与数据库交互接口 UserMapper，在接口中声明访问数据库的方法。

```
package com.example.dao;

import java.util.List;
import java.util.Map;
import org.apache.ibatis.annotations.Mapper;
import com.example.entity.UserSearchDTO;
import com.example.pojo.User;

@Mapper
public interface  UserMapper {

    /**
     * 分页查询用户数据
     * @return
     */
```

```
    List<User> getUsers(UserSearchDTO userSearch);

    User queryById(String id);

    int editUser(User user);

    int insertUser(User user);

    int delUser(String[] ids);

    int delRoles(Map<String,Object> map);

    int addRoles(Map<String,Object> map);

    User getUserById(String id);
}
```

项目使用基于 xml 的 mybatis 来操作数据库，上面定义了接口之后，还需要在 resource 下的 mybatis 目录下创建 sql 映射文件 UserMapper.xml，映射文件已经存在，但项目还不知道数据库地址、sql 映射文件位置等信息，所以还需要在 application.properties 中设置数据库信息和 sql 映射文件位置等信息。

```
#mysql
spring.datasource.driverClassName = com.mysql.cj.jdbc.Driver
#spring.datasource.url = jdbc:mysql://localhost:3306/mybatis
spring.datasource.url =jdbc:mysql://127.0.0.1:3306/mybatis?useUnicode=
true&characterEncoding=UTF-8&serverTimezone=UTC
spring.datasource.username = root
spring.datasource.password = 123456
#mybatis
mybatis.type-aliases-package=com.example.pojo
mybatis.config-location=classpath:mybatis/mybatis-config.xml
mybatis.mapper-locations=classpath:mybatis/mapper/*.xml
```

访问层配置完之后，配置服务层。在 service 接口层定义接口，然后在 service.impl 层实现该接口。下面的代码是 service 接口层代码，具体实现层代码可以查看项目代码。

```
package com.example.service;
import org.springframework.stereotype.Service;

import com.example.entity.UserSearchDTO;
import com.example.pojo.User;
import com.example.utils.PageDataResult;

@Service
public interface UserService {
```

```
    public PageDataResult getUsers(UserSearchDTO userSearch);

    public User queryById(String id);

    public Boolean editUser(User user);

    public Boolean insertUser(User user);

    public Boolean delUser(String id) throws Exception;

    public Boolean delRoles(String userId,String roleIds);

    public Boolean addRoles(String userId,String roleIds);

    public User getUserById(String id);
}
```

服务层配置完之后就可以在 Controller 中注入服务层的 service，调用服务层的方法了。用户操作页面发送请求调用 Controller 中的方法。

3. 分页

layui 的 table 填充的数据有 4 个字段，即 code、msg、count、data，所以在 PageDataResult 中也定义了 4 个属性。

```
package com.example.utils;

import java.util.List;

public class PageDataResult {
    //总记录数量
    private Integer totals;
    //当前页数据列表
    private List<?> list;

    private Integer code=200;

    private String msg="";

    public String getMsg() {
        return msg;
    }

    public void setMsg(String msg) {
        this.msg = msg;
    }

    public PageDataResult() {
```

```
    }

    public PageDataResult( Integer totals,
            List<?> list) {
        this.totals = totals;
        this.list = list;
    }

    public Integer getTotals() {
        return totals;
    }

    public void setTotals(Integer totals) {
        this.totals = totals;
    }

    public List<?> getList() {
        return list;
    }

    public void setList(List<?> list) {
        this.list = list;
    }

    public Integer getCode() {
        return code;
    }

    public void setCode(Integer code) {
        this.code = code;
    }

    @Override public String toString() {
        return "PageDataResult{" + "totals=" + totals + ", list=" + list
                + ", code=" + code + '}';
    }
}
```

使用 pagehelper 插件分页之后的数据拼装成 PageDataResult 对象返回前端页面。

```
@Override
public PageDataResult getUsers(UserSearchDTO userSearch) {

        PageDataResult pdr = new PageDataResult();
        PageHelper.startPage(userSearch.getPage(),
userSearch.getLimit(),true);
        List<User> urList = userMapper.getUsers(userSearch);
```

```
            logger.debug("urList:"+urList.size());
            // 获取分页查询后的数据
            PageInfo<User> pageInfo = new PageInfo<>(urList);
            // 设置获取到的总记录数 total:
            logger.debug("page:"+userSearch.getPage()+"limit:"+
userSearch.getLimit()+"总行数:"+pageInfo.getTotal());
            pdr.setTotals(Long.valueOf(pageInfo.getTotal()).intValue());
            pdr.setList(urList);
            return pdr;

    }
```

由于与 LayUI 的 table 模块要求的数据字段名不一致，可以使用 response 属性配置，而且列表性别列中，数据返回的是 0、1，为了得到更好的用户体验，使用 templet 属性将 1 显示为男、0 显示为女。

```
    layui.use([ 'layer', 'table', 'element' ], function() {
       table = layui.table;
       layer = layui.layer;
       // 执行一个 table 实例
       table.render({
          elem : '#user',
          height:350,
          url : '/user/getUsers',
          method: 'get', //默认: get 请求
          page :true, // 开启分页
          request: {
             pageName: 'page' //页码的参数名称，默认: page
             ,limitName: 'limit' //每页数据量的参数名，默认: limit
          },response:{
             statusName: 'code' //数据状态的字段名称，默认: code
             ,statusCode: 200 //成功的状态码，默认: 0
             ,countName: 'totals' //数据总数的字段名称，默认: count
             ,dataName: 'list' //数据列表的字段名称，默认: data
          },
          cols : [ [ // 表头
             {
                fixed : 'left',
                type : 'checkbox'
             },{
                field : 'id',
                title : '编号',
                width : 80
             }
             , {
                field : 'name',
                title : '姓名',
```

```
                width : 80
            },
            {
                field : 'age',
                title : '年龄',
                width : 80
            },
            {
                field : 'sex',
                title : '性别',
                width : 80,
                templet : function(d) {
                    if (d.sex == 1) {
                        return '男';
                    } else if (d.sex == 0) {
                        return '女';
                    }
                }
            },{
                title : '操作',
                width : 200,
                align : 'center',
                toolbar : '#tools'
            } ] ]

    });
```

4. 主子表维护

对 User 表的操作属于单表操作，实现了 User 单表操作之后角色 Role 表、权限 Permission 表的增删改查操作可以参考用户 User 表实现。项目中常见的不仅有单表操作，还有主子表操作。用户和角色有一个关系表 user_role 表，对主子表的维护其实也是对关系表的操作。

用户在用户表中点击“选择角色”（见图 11-6），弹出弹窗显示已有的角色，点击“添加角色”时，在弹窗上面再弹出一个窗体显示用户未关联的角色。用户可以选择角色，点击“添加”按钮（见图 11-7）之后，在顶部弹窗会将选中的角色去除，选择完毕关闭顶部弹窗时会刷新底部弹窗，显示用户的角色，同时用户可以对关联角色进行删除，具体实现可以参考项目代码。

5. 事务处理

主子表带来一个问题，就是当删除主表数据时需要把子表关联的数据也删除，比如用户页面，删除用户就需要把用户与角色关联表中相应的用户数据也删除。在 Spring Boot 处理事务，只需要在 main 方法类中开启全局事务，然后在对应的类或方法上声明事务即可。

图 11-6

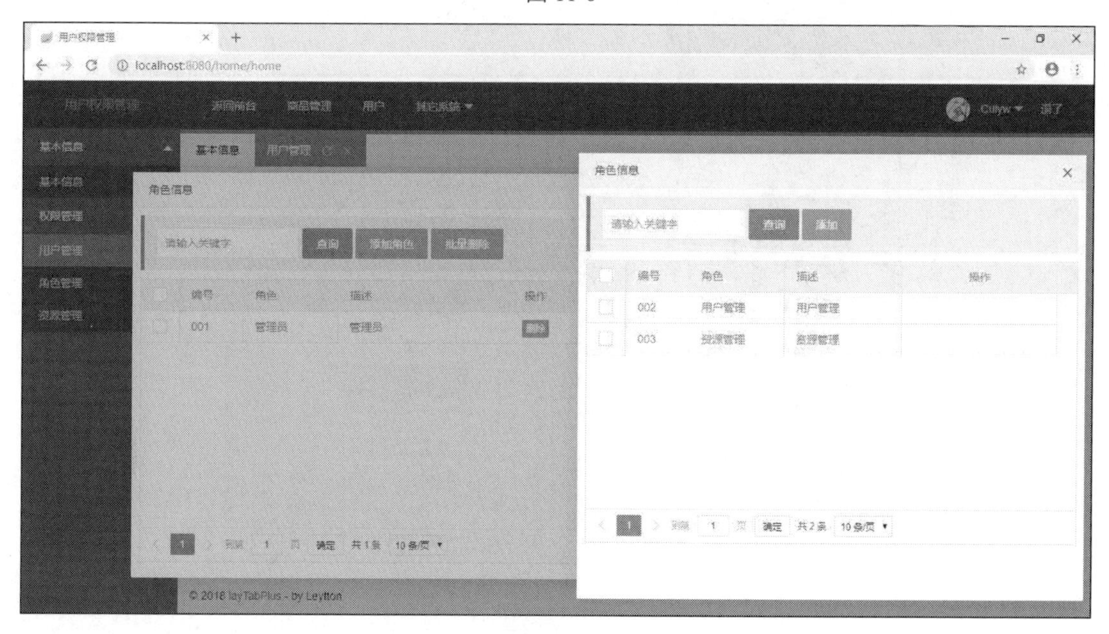

图 11-7

（1）开启事务

在 main 方法类中使用@EnableTransactionManagement 注解开启事务。

```
@ComponentScan(basePackages = {"com.example.*"})
@SpringBootApplication
@EnableTransactionManagement
public class ManageApplication {

    public static void main(String[] args) {
        SpringApplication.run(ManageApplication.class, args);
    }


}
```

（2）在类或方法上声明事务

使用@Transactional 注解在类、方法上声明事务。删除用户的时候需要删除关联表中的角色信息，在删除用户的方法上声明了事务，为了测试事务，在删除关联表时抛出了 RuntimeException 异常。用户编号 001 的用户关联的是有角色的，在点击“删除”时会报“删除失败！”，如图 11-8、图 11-9 所示。

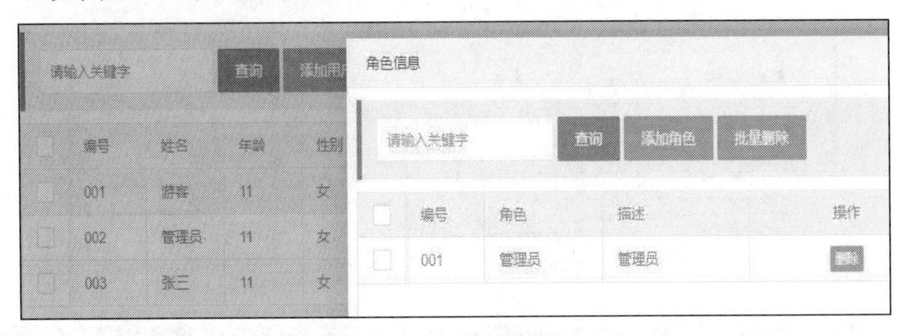

图 11-8

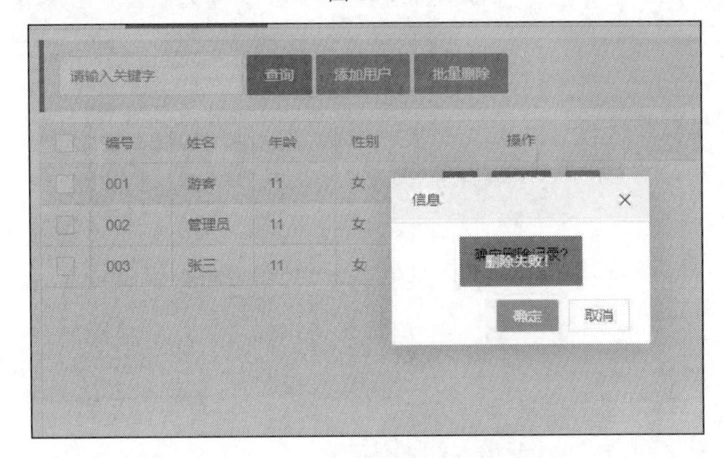

图 11-9

```
@Override
@Transactional(propagation=Propagation.REQUIRED)
public Boolean delUser(String id) throws Exception {
    String[]  ids=id.split(",");
    if(ids.length>0)
    {
        userMapper.delUser(ids);
        for(int i=0;i<ids.length;i++)
        {
            delRoles(ids[i],"");
            throw new RuntimeException("注释");
        }
    }
    return true;
}
```

11.2.5　Shiro 用户权限管理

用户权限管理一般是对用户页面、按钮的访问权限管理。本小节主要介绍 Shiro 的简单使用。

1. 引入依赖

使用 Spring Boot 集成 Shiro 时，在 pom.xml 中可以引入 shiro-spring-boot-web-starter。这里使用的是 Thymeleaf 框架，Thymeleaf 与 Shiro 结合需要引入 thymeleaf-extras-shiro。

```
<!-- https://mvnrepository.com/artifact/org.apache.shiro/
shiro-spring-boot-web-starter -->
<dependency>
    <groupId>org.apache.shiro</groupId>
    <artifactId>shiro-spring-boot-web-starter</artifactId>
    <version>1.4.0</version>
</dependency>
<!-- https://mvnrepository.com/artifact/
com.github.theborakompanioni/thymeleaf-extras-shiro -->
<dependency>
    <groupId>com.github.theborakompanioni</groupId>
    <artifactId>thymeleaf-extras-shiro</artifactId>
    <version>2.0.0</version>
</dependency>
```

2. 增加 Shiro 配置

有哪些 url 是需要拦截的、哪些是不需要拦截的以及登录页面、登录成功页面的 url、自定义的 Realm 等这些信息需要设置到 Shiro 中，所以创建 Configuration 文件 ShiroConfig。其中，ShiroDialect bean 对象是在 Thymeleaf 与 Shiro 结合、前端 HTML 访问 Shiro 时使用的。

```
@Configuration
public class ShiroConfig {
    @Bean("shiroFilterFactoryBean")
    public ShiroFilterFactoryBean shiroFilterFactoryBean(SecurityManager
securityManager) {
        System.out.println("ShiroConfiguration.shirFilter()");
        ShiroFilterFactoryBean shiroFilterFactoryBean = new
ShiroFilterFactoryBean();
        shiroFilterFactoryBean.setSecurityManager(securityManager);
        //拦截器.
        Map<String,String> filterChainDefinitionMap = new LinkedHashMap
<String,String>();
        // 配置不会被拦截的链接 顺序判断
        filterChainDefinitionMap.put("/static/**", "anon");
        //配置退出过滤器,其中的具体退出代码 Shiro 已经替我们实现了
        filterChainDefinitionMap.put("/logout", "logout");
```

```
        //<!-- 过滤链定义，从上向下顺序执行，一般将/**放在最为下边 -->:这是一个坑呢，
一不小心代码就不好使了;
        //<!-- authc:所有 url 都必须认证通过才可以访问; anon:所有 url 都可以匿名访问
-->
        filterChainDefinitionMap.put("/**", "authc");
        // 如果不设置默认会自动寻找 Web 工程根目录下的"/login.jsp"页面
        shiroFilterFactoryBean.setLoginUrl("/login");
        // 登录成功后要跳转的链接
        shiroFilterFactoryBean.setSuccessUrl("/index");
        //未授权界面;
        shiroFilterFactoryBean.setUnauthorizedUrl("/403");
        shiroFilterFactoryBean.setFilterChainDefinitionMap
(filterChainDefinitionMap);
        return shiroFilterFactoryBean;
    }

    //创建 defaultWebSecurityManager
    @Bean(name="defaultWebSecurityManager")
    public DefaultWebSecurityManager getDefaultWebSecurityManager
(@Qualifier("userRealm")MyShiroRealm userRealm){
        DefaultWebSecurityManager defaultWebSecurityManager = new
DefaultWebSecurityManager();
        defaultWebSecurityManager.setRealm(userRealm);
        return defaultWebSecurityManager;

    }

    //创建 Realm
    @Bean(name="userRealm")
    public MyShiroRealm getUserRealm(){
        return new MyShiroRealm();
    }

    @Bean
  public ShiroDialect shiroDialect() {
     return new ShiroDialect();
   }
  }
```

3. 自定义 Realm

在自定义的 Realm 中继承了 AuthorizingRealm 抽象类，重写了两个方法：doGetAuthorizationInfo 和 doGetAuthenticationInfo。doGetAuthorizationInfo 主要用来处理权限配置，doGetAuthenticationInfo 主要处理身份认证。在 doGetAuthorizationInfo 中，将 role 表的 id 和 permission 表的 code 分别设置到 SimpleAuthorizationInfo 对象中的 role 和 permission 中。最后需要在自定义的 realm 上添加@Component("authorizer")，否则会报错，提示需要一个 authorizer 的 bean。

```
    @Component("authorizer")
    public class MyShiroRealm extends AuthorizingRealm {

        @Autowired
        private UserService userService;

        @Autowired
        private RoleService roleService;

        @Override
        protected AuthorizationInfo doGetAuthorizationInfo(PrincipalCollection
principals) {
            System.out.println("权限配置-->MyShiroRealm.
doGetAuthorizationInfo()");
            SimpleAuthorizationInfo authorizationInfo = new
SimpleAuthorizationInfo();
            User user  = (User)principals.getPrimaryPrincipal();
            System.out.println("User:"+user.toString()+" roles count:"+
user.getRoles().size());
            for(Role role:user.getRoles()){
                authorizationInfo.addRole(role.getId());
                role=roleService.getRoleById(role.getId());
                System.out.println("Role:"+role.toString());
                for(Permission p:role.getPermissions()){
                System.out.println("Permission:"+p.toString());
                    authorizationInfo.addStringPermission(p.getCode());
                }
            }
            System.out.println("权限配置-->authorizationInfo"+
authorizationInfo.toString());
            return authorizationInfo;
        }

        /*主要是用来进行身份认证的，也就是说验证用户输入的账号和密码是否正确。*/
        @Override
        protected AuthenticationInfo doGetAuthenticationInfo
(AuthenticationToken token)
                throws AuthenticationException {
            System.out.println("MyShiroRealm.doGetAuthenticationInfo()");
            //获取用户的输入账号
            String username = (String)token.getPrincipal();
            System.out.println(token.getCredentials());
            //实际项目中，这里可以根据实际情况做缓存，如果不做，Shiro 自己也是有时间间隔机
制的，2 分钟内不会重复执行该方法
            User user = userService.getUserById(username);
            System.out.println("----->>userInfo="+user);
            if(user == null){
                return null;
            }
```

```
            SimpleAuthenticationInfo authenticationInfo = new
 SimpleAuthenticationInfo(
                user, //用户名
                "123456", //密码
                  getName()  //realm name
            );
            return authenticationInfo;
        }
    }
```

4. 登录认证

（1）登录页面

在 templates 目录下定义了 login.html，用来作为登录页面。页面也比较简单，在表单中输入用户名、密码，然后提交表单即可。

```
<form action="/login" method="post">
    <label>用户名：</label><input type="text" name="id"  id="id" ><br>
    <label >密码：</label><input type="text" name="pwd"  id="pwd" ><br>
    <button type="submit">登录</button><button type="reset">取消</button>
</form>
```

（2）处理登录请求

在 LoginController 中通过登录名、密码获取到 token 实现登录。

```
@RequestMapping(value = "/login",method = RequestMethod.POST)
public String login(Model model,String id,String pwd){
    //添加用户认证信息
    Subject subject = SecurityUtils.getSubject();
    UsernamePasswordToken usernamePasswordToken = new
 UsernamePasswordToken(
             id,
             "123456");
    try {
        subject.login(usernamePasswordToken);
        return "home";
    }
    catch (UnknownAccountException e) {
        //用户名不存在
        model.addAttribute("msg","用户名不存在");
        return "login";
    }catch (IncorrectCredentialsException e) {
        //密码错误
        model.addAttribute("msg","密码错误");
        return "login";
    }

}
```

5. Controller 层访问控制

（1）首先看一下数据库的数据，用户角色、角色权限的数据如图 11-10、图 11-11 所示。

Result Grid | Filter Rows: | Export: | Wrap Cell Content:

user_id	user_name	user_age	user_sex	role_id	role_name	role_desp
001	游客	11	0	001	管理员	管理员
002	管理员	11	0	001	管理员	管理员
003	张二	11	0	001	管理员	管理员
002	管理员	11	0	002	用户管理	用户管理
002	管理员	11	0	003	资源管理	资源管理

图 11-10

Result Grid | Filter Rows: | Export: | Wrap Cell Content:

role_id	role_name	role_desp	permission_id	permission_name	permission_code	permission_desp	permission_url
001	管理员	管理员	6	管理员	001	管理员	/user/all
002	用户管理	用户管理	6	管理员	001	管理员	/user/all
003	资源管理	资源管理	NULL	NULL	NULL	NULL	NULL

图 11-11

（2）设置权限。控制 url 访问权限可以在 Controller 中的方法上面使用@RequiresRoles 或者@RequiresPermissions 注解。

```
@RequestMapping(value = "/edit",method = RequestMethod.GET)
@RequiresRoles("002")//权限管理;
public String editGet(Model model,@RequestParam(value="id") String id){
    model.addAttribute("id", id);
    return "/user/edit";
}

@RequestMapping(value = "/selrole",method = RequestMethod.GET)
@RequiresPermissions("002")//权限管理;
public String selctRole(Model model,@RequestParam("id") String id,
@RequestParam("type") Integer type) {
    model.addAttribute("id",id);
    model.addAttribute("type", type);
      return "/user/selrole";
}
```

上面在用户页面点击“编辑”按钮时设置需要有 id=002 的角色，在点击“选择角色”按钮时需要有 code=002 的权限。由于 001 用户只有 001 的角色和 001 的权限，因此当使用用户 001 登录时，点击“编辑”，弹出框如图 11-12 所示，提示没有 002 的角色，点击“选择角色”按钮时，提示没有 002 的权限，如图 11-13 所示。当使用用户 002 登录时，用户 002 有 001、002、003 三种角色，所以点击“编辑”按钮时显示正常，点击“选择角色”时也是提示没有 002 的权限，因为权限只有 001。

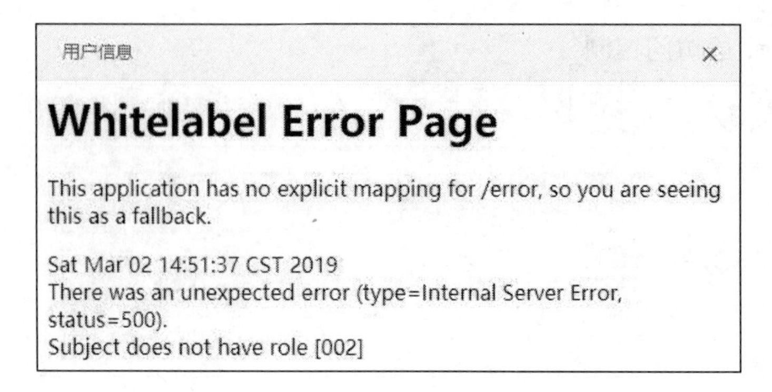

图 11-12

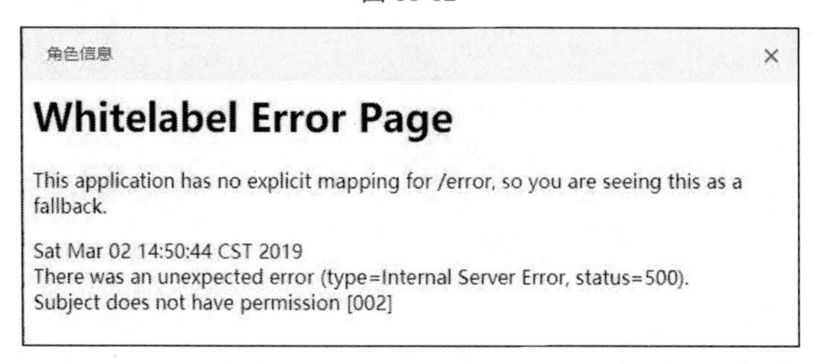

图 11-13

6. 前端页面层访问控制

为了不像上面那样弹出错误页面，需要在按钮显示上设置不可见，这样用户也不会点击到。在前端使用 Shiro 需要引入依赖 thymeleaf-extras-shiro，前面已有引入。同时还需要在前端页面中设置 html 标签引入 Shiro。

```
<html xmlns:th="http://www.thymeleaf.org"
      xmlns:shiro="http://www.pollix.at/thymeleaf/shiro">
```

为了测试页面按钮，使用 shiro:hasAnyRoles="002,003"判断用户角色是否是 002 或 003，是就显示，不是则不显示添加用户、批量删除按钮。

```
<div class="layui-inline">
    <a shiro:hasAnyRoles="002,003" class="layui-btn layui-btn-normal
newsAdd_btn" onclick="addUser('')">添加用户</a>
</div>
<div class="layui-inline">
    <a shiro:hasAnyRoles="002,003" class="layui-btn layui-btn-danger
batchDel" onclick="getDatas();">批量删除</a>
</div>
```

当 001 用户登录时，由于没有 002、003 的角色，因此添加用户、批量删除按钮都不显示，只显示查询按钮，如图 11-14 所示。

图 11-14

当 002 用户登录时，由于有 001、002、003 三种角色，因此添加用户、批量删除按钮都显示，如图 11-15 所示。

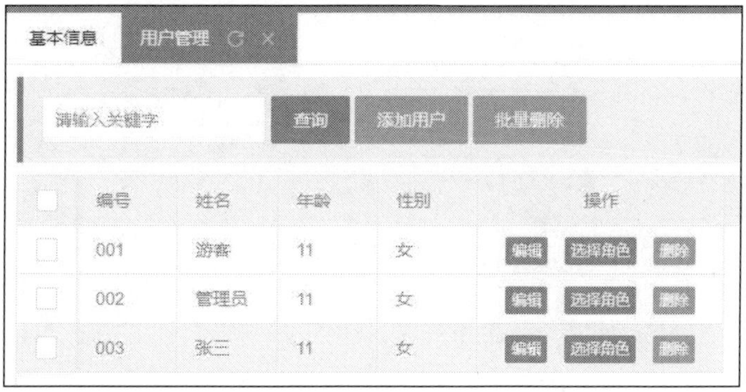

图 11-15

11.3 小结

该项目只是实现了 Shiro 的简单功能，其实 Shiro 还有很多很强大的功能，比如 Session 管理等，而且权限管理模块还有很多需要优化的功能，左侧导航栏的动态加载和权限控制、Shiro 与 Redis 结合实现 Session 共享、Shiro 与 Cas 结合实现单点登录等。后续还可以把项目开源，集成更多模块供初学者参考。